The Effects of Hydrogen in Aluminium and its Alloys

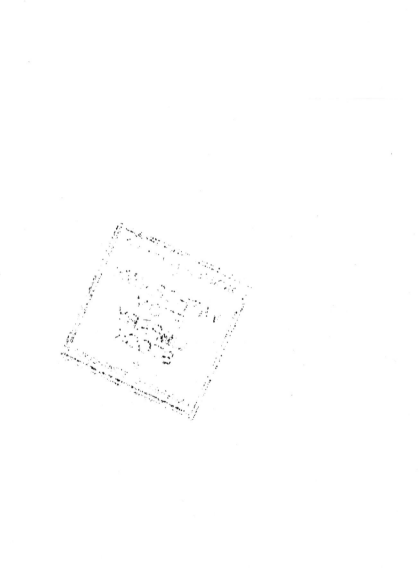

The Effects of Hydrogen in Aluminium and its Alloys

D.E.J. Talbot

MANEY

FOR THE INSTITUTE OF MATERIALS, MINERALS AND MINING

B0724
First published in 2004 for
The Institute of Materials, Minerals and Mining by
Maney Publishing
1 Carlton House Terrace
London SW1Y 5DB

Maney Publishing is the trading name of
W. S. Maney and Son Ltd
Hudson Road
Leeds LS9 7DL

ISBN 1-902653-73-4

Typeset in India by Emptek Inc.
Printed and bound in the UK by
The Charlesworth Group

Contents

6. SOLUBILITIES OF HYDROGEN IN ALUMINIUM AND ITS ALLOYS

PREFACE

Interpretation of information from industrial observations on interactions between gases and metals that can influence the quality and yield of manufactured products requires inputs from a wide spectrum of underlying physical science. The purpose of this text is to integrate these mutually supportive aspects relevant to the effects hydrogen on aluminium and its alloys.

Industrial interest is driven by economic imperatives to satisfy standards for quality and yield of metallic products. This is especially true of interactions between hydrogen and aluminium alloys that are typically produced for critical applications which are highly sensitive to damage by hydrogen absorbed during manufacture. As quality standards rise ever higher, experience accumulated from observation on metal in production and related experiments on an industrial scale have yielded elegant integrated systems for liquid metal treatment, metal transfer, casting, fabrication and heat treatment facilities.

Theoretical interest is stimulated by the unique characteristics of the hydrogen atom by virtue of its small size and the single electron in the is shell conferring on it the facility to form interstitial solutions in metals and to transform between molecules, neutral atoms, anions and protons. These fundamental features explain many practical aspects of its interactions with aluminium and other metals, e. g. the forms in which it exists within metal structures and the reaction kinetics by which it enters the metal from environmental sources.

Chapters 1 to 3 establish the context by an overview and brief summaries of the constitutions of aluminium alloys and of manufacturing processes. Chapter 4 derives the scientific infrastructure underlying the occlusion of hydrogen. Chapters 5, 6 and 7 describe the measurement of hydrogen contents, solubilities and diffusivities of hydrogen in liquid and solid metal and critically assess the reliability and significance of values obtained. Chapter 8 explains the absorption of hydrogen from the environment in terms of the constitutions of oxidation products and the capacity of the metal to receive the gas. Chapters 9 and 10 describe the incidence and control of artifacts.

ACKNOWLEDGEMENTS

This monograph is dedicated to Dr C. E. Ransley, in recognition of his pre eminence in the field of gas/metal systems by introducing the theoretical approaches and innovative measurement systems now applied as standard for the resolution of industrial problems.

I am indebted to Dr Douglas Granger for the benefit of his unrivalled experience of quality issues in the Aluminium Industry and to Professor Colin Bodsworth and Professor Brian Ralph for their encouragement and strict but always constructive criticism. It is a pleasure to express my appreciation of the skilful experimental work, intellect and dedication of former postgraduate students at Brunel University, especially David Stephenson, Princewill Anyalebechi, M. Prem Silva, Margaret Sargent, Joanne Morton, Colin McCracken, Dilys Henry and Masood Al Rais.

Material on the constitution of aluminium alloys in Chapter 2 is taken from an earlier book *Corrosion Science and Technology* by kind permission of CRC Press.

1. Overview of the Effects of Occluded Hydrogen

Most common engineering metals can dissolve one or more of the elemental gases hydrogen, oxygen and nitrogen from the environment in the course of manufacture. The active species vary from metal to metal depending on the solubilities of the gases and the potential for interaction with other solutes to form compound gases, e.g. carbon monoxide and sulphur dioxide. Characteristic artifacts and effects include porosity due to the evolution of carbon monoxide during solidification of incompletely deoxidised steels, hydrogen embrittlement of steel forgings and of some nickel base alloys, nitrogen embrittlement of ferritic steels, evolution of sulphur dioxide and steam during the solidification of tough pitch copper and steam embrittlement of oxygen–bearing copper annealed in atmospheres containing hydrogen.

Nitrogen, oxygen, sulphur and carbon are all insoluble in aluminium and its alloys, so that hydrogen is the only gaseous species for consideration. The quantities that are significant are much less than quantities that have little effect in other metals because the solubility in solid aluminium is so low that hydrogen absorbed in the course of manufacture does not remain in solution but is almost invariably partitioned between solution and various traps. Without control of the hydrogen content and thermal treatment of the metal, these traps can develop into artifacts which so impair the bulk properties and surface qualities of the products that they are unsuitable for critical applications that form a large part of the market for the metal.

Aluminium is second only to steel among economically important metals. It is relatively expensive but it has a remarkable combination of qualities that includes low density, high ductility, high thermal and electrical conductivity, good corrosion resistance, attractive appearance, and non-toxicity. These attributes make the pure metal and alloys based on it preferred choices for applications in aerospace, automobiles, food handling, building, heat exchange and electrical transmission. Many of these applications are critical and impose stringent requirements for freedom from artifacts that can compromise the integrity or appearance of the metal, usually described by the term *metal quality*.

Industrial processes yielding aluminium products are normally conducted in environments containing sources of hydrogen from which they absorb the gas as a contaminant. These products are vulnerable to damage by the hydrogen they absorb because the solubility of the gas in the solid metal is low. The damage is due to internal or subsurface defects that are essentially extended hydrogen traps of one form or another, some inherited from unsoundness and inclusions in castings and ingots and others introduced during subsequent thermal and mechanical treatment of the solid metal in course of fabrication to semi-finished products. Characteristic examples of the damage[1] are blistering or other surface blemishes on heat-treated rolled sheet and extrusions, discontinuities across load-bearing sections, inadequate surface finish for anodic quality products and failure to contain fluids in near-net shape castings. These effects once took a heavy economic toll in lost quality and reduced yield. Moreover the recovery of useful metal was unpredictable, incurring costs for make up charges and unwanted excesses. The incidence of defects related to hydrogen is now quite low due to the replacement of empiricism by technologies described in the present text.

Chapters 2 and 3 are preliminaries briefly introducing the range of aluminium and its alloys and the processes by which products are manufactured from them. Chapter 4 considers the states in which hydrogen is occluded in the metal. Chapters 5, 6 and 7 describe the specialised techniques required to measure hydrogen contents, solubilities and diffusion coefficients. Chapter 8 deals with the absorption of hydrogen by the metal from environmental sources and Chapters 9 and 10 describe the incidence and control of artifacts associated with hydrogen.

1.1 STATES OF OCCLUSION

Hydrogen occluded in aluminium and its alloys is usually distributed between solution and traps of various kinds and it is often interaction between the solution and traps that is the genesis of defects in the metal.

1.1.1 NATURE OF HYDROGEN

The hydrogen atom has unique characteristics by virtue of its small atomic size and the single electron in the 1s shell conferring on it the facility to transform easily between molecules, neutral atoms, anions, or protons. Protons are very small with high charge densities in relation to other atoms and in condensed matter they are invariably associated with other atoms or molecules. These features are of special interest in explaining interactions of hydrogen with metals, e. g. the forms in which it can exist within metal structures. The composition of natural hydrogen is > 99.99% of the isotope, ^{1}H. In the present context, the other isotopes, deuterium, ^{2}H, and tritium, ^{3}H are of interest only to represent hydrogen experimentally in neutron scattering and tracer techniques.

Molecular hydrogen is diatomic and its dissociation is highly endothermic:[2]

$$H_2 = 2H \qquad\qquad (1.1)$$

for which:

$$\Delta H^{\circ}(298) = +427 \text{ kJ} \tag{1.2}$$

1.1.2 SOLUTIONS

Information on solutions of hydrogen can be gathered by simple experimental methods because the gas is virtually ideal at elevated temperatures and the solutions are so dilute that Henrian activity applies to the solute. Hence solubilities and diffusivities can be obtained directly from the results of isothermal sorption and desorption. The concentration of hydrogen dissolved in the metal in equilibrium with the diatomic gas is a parabolic function of the gas pressure,[3, 4] indicating that the solute is present as a monatomic species with minimal self interaction.

All of the approaches given in Chapter 4 are required to characterise the solutions. Classical thermodynamics and statistical formulation yield equations for the pressure and temperature dependence of solutions, expressed as Sievert's isotherm and Van't Hoff's isobar. The enthalpy of solution can be determined by treating the bond between the solute and the metal as an interaction between solute atoms and the prevailing electron density,[5] providing information on monatomic traps at lattice defects and interfaces. The entropy of solution is computed from the sum of the entropies of dissociation of the diatomic gas, the change of partition function and mixing among the octahedral interstices; this sum is identical for interstitial solutions of hydrogen in all FCC metals and the information is required to interpolate experimentally determined solubilities of hydrogen in pure aluminium and to identify preferred sites for hydrogen in aluminium-lithium alloys.

The concentration of hydrogen in solution is limited to that in equilibrium with the excess pressure due to the surface tension of prevailing pore nuclei, which are ubiquitous in manufactured products. This leads to spontaneous breakdown of solutions by heterogeneous nucleation and growth of micropores.

1.1.3 HYDROGEN TRAPS

In general, occluded hydrogen is distributed between solution and molecular traps. One kind of traps is interdendritic porosity in ingots cast from inadequately prepared liquid metal and its residues in wrought forms fabricated from them. Another kind is microporosity containing molecular hydrogen nucleated in the solid metal from the spontaneous breakdown of solutions referred to in Section 1.1.2. Chemical trapping is not usually a consideration for aluminium and its alloys since neither the pure metal nor the common alloying elements form stable hydrides under conditions of metallurgical interest. It is significant only in the special case of alloys containing lithium.

1.2 QUANTITATIVE APPROACHES

1.2.1 HYDROGEN CONTENTS

Reliable methods for determining quantities of occluded hydrogen in aluminium and its alloys,[6, 7] described in Chapter 5, are responsible for most of the progress that has been made in characterising the behaviour of hydrogen in manufactured aluminium and aluminium alloy products. Two complementary methods have stood the test of time, vacuum extraction in which hydrogen evolved from a prepared sample is collected and measured and the Telegas instrument that gives instant values for the activity of hydrogen in liquid metal. Each has its particular role. There is no alternative to vacuum extraction analysis for solid metal. Although it can be used for samples cast from liquid metal, it is slow and delivers analyses retrospectively. The Telegas is convenient to monitor the hydrogen content of liquid metal where it can be accommodated, as in a transfer launder conveying liquid metal from a furnace to ingot moulds. Both techniques require dedicated operators experienced in the sources of error to which they are vulnerable.

1.2.2 SOLUBILITIES

Values for the solubilities of hydrogen in solid and liquid aluminium and its alloys are required as functions of composition, temperature and pressure to serve several purposes:
1.   to characterise the natures of the solutions,
2.   for comparison with the prevailing hydrogen contents of metal in production,
3.   to assess activities of dissolved hydrogen and
4.   to calibrate the Telegas instrument for hydrogen content determinations.
    Solubilities of hydrogen in liquid metal are usually determined by *Sieverts' method* in which isothermal isobaric absorption of the gas by a prepared sample is measured volumetrically. It is less suitable for solid metal because the slow diffusion-controlled sorption of the gas introduces time-dependent sources of error. A more suitable method is to measure the hydrogen contents of samples by vacuum extraction after equilibrating them with hydrogen. Both methods are described in detail and reported values of solubility are critically assessed, tabulated and given as equations for reference in Chapter 6.
    The solubilities of hydrogen in pure aluminium and in binary alloys with copper and silicon conform with Sievert's isotherm and the Van't Hoff isobar as expected for endothermic solutions.[8] A significant feature is the much higher ratio of the solubilities in liquid and solid aluminium than for other metals.
    The solubilities of hydrogen in liquid and solid alloys containing lithium are much higher than in other alloys and every isobar for solid alloys is in two parts separated by a discontinuity at a critical temperature that depends on composition.[9] Comparison with the corresponding isobar for pure solid aluminium provides evidence that the high solubilities are due to enhanced accommodation for the solute at preferred sites. These

sites have been identified as lithium-rich clusters from evidence obtained by small angle neutron diffraction[10, 11] that dissolved hydrogen delays the ripening of clusters preceding the precipitation of δ' during ageing.

1.2.3 DIFFUSIVITY

Values for the diffusivities of hydrogen in aluminium and its alloys are needed to assess rates at which it can be absorbed and distributed in the metal during thermal treatments. The true diffusivity of hydrogen in solid pure aluminium is of only academic interest because it applies to hydrogen exclusively in solution. It has been determined from the desorption kinetics for laboratory samples prepared by prolonged annealing in vacuum to ensure that the metal is completely sound before charging it with hydrogen.[3]

In manufactured aluminium and aluminium alloy products, occluded hydrogen is distributed between the monatomic species in solution and molecular hydrogen in disseminated microporosity,[12] as described in Section 1.1.3, so that the transport of hydrogen is by diffusion in a field of traps. This diffusing system is described by standard diffusion equations modified by introducing a non-linear isotherm describing local equilibrium between the monatomic species in solution and molecular hydrogen in the traps.[13] With certain reasonable assumptions, manageable equations are obtained that are of the same form as standard diffusion equations, where the diffusion coefficient, D, is replaced by an effective coefficient, D'. This approach describes the transport of hydrogen in manufactured products very well. Transport of hydrogen in aluminium–lithium alloys is also treated as diffusion in a field of traps with the difference that local equilibrium between the free solute and solute trapped in the preferred sites is described by a linear isotherm.

Values for diffusion coefficients and effective diffusion coefficients are tabulated for convenient reference in Tables 7.3-7.5.

1.3 ENVIRONMENTAL SOURCES OF HYDROGEN

1.3.1 HUMIDITY

Metal in production can absorb hydrogen from humid air, hydrated materials with which it is in contact and recycled scrap. The standard Gibbs free energy changes[14] for the reactions:

$$2Al(\text{solid or liquid}) + 3H_2O(\text{gas}) = Al_2O_3(\text{solid}) + 3H_2(\text{gas}) \tag{1.3}$$

$$\Delta G^{\ominus} = -958387 - 72.87T \log T + 394.5T \text{ J (for solid aluminium)} \tag{1.4}$$

and

$$\Delta G^{\oplus} = -979098 - 71.91T \log T + 413.6T \text{ J (for liquid aluminium)} \qquad (1.5)$$

are so high that virtually all traces of water contacting the metal are converted to hydrogen. Water and its vapour cannot be completely eliminated from production environments, so that the factors that control absorption of hydrogen by aluminium products are reaction kinetics and the capacity of the metal to receive the gas.

Reaction between pure aluminium and humid air at elevated temperatures produces oxidation products on the metal surface that separate the metal from the atmosphere so that reaction can continue only by transport of reacting species through the oxidation products. The constitutions of these products[15] is determined by the water vapour potentials at the oxide/metal and atmosphere/oxide air interfaces. The initial oxidation product is a water stabilised transition alumina, η-Al_2O_3 and an underlayer of corundum nucleates and grows at the oxide/metal interface. Before the corundum layer is fully established the metal absorbs hydrogen by reaction with hydroxyl ions transported through the η-Al_2O_3 but when the corundum layer is continuous across the whole oxide/metal interface the reaction ceases because hydroxyl ions are structurally incompatible with corundum. This means that liquid aluminium can absorb hydrogen from atmospheric water vapour but only for a limited period. Solid aluminium does not readily absorb hydrogen from humid air because the continuous layer of corundum is established before much hydrogen can diffuse into the metal.

The oxidation product on solid alloys containing magnesium in excess of 1 mass % is magnesium oxide,[16] MgO which is stable with respect to magnesium hydroxide but it can accept a few hydroxyl ions as impurities. This allows the metal to absorb hydrogen slowly from clean humid air at elevated temperatures but only to the extent that it can be accommodated in pre-existing microporosity with no significant damage to the metal. The interaction is accelerated by a trace of sulphur dioxide in the atmosphere, which modifies the oxidation product, allowing hydrogen to be absorbed so strongly that inflation of pore nuclei can render the metal unserviceable. The absorption can be inhibited in industrial furnaces by introducing small quantities of solid potassium borofluoride or sodium silicofluoride[17] as a source of a volatile fluoride that forms a protective film on the metal surface.

The oxidation product on alloys containing lithium heated in clean humid air is lithium hydroxide which can freely conduct hydroxyl ions to the oxide/metal interface and the metal rapidly absorbs hydrogen as lithium hydride nucleated just below the metal surface.

1.3.2 ROLLED SURFACES

Sheet and plate products are usually manufactured by hot-rolling preheated ingots with multiple passes to an intermediate material with a thickness in the range 4 to 12 mm, which is then cold-rolled to the final gauge if required. Relative motion between the metal and the roll face implants detritus and rolling lubricant residues in the metal surface. This condition maintains a constant hydrogen potential at the metal surface during subsequent heating that drives hydrogen into the metal at a rate controlled by diffusion which becomes significant at temperatures above 400°C. Cold rolling further enhances the effect of the surface source by increasing the surface/volume ratio and adding further contamination from trapped lubricant. A theoretical model simulating a typical thermal cycle in industrial annealing and based on the effective diffusion coefficient represents the rate at which hydrogen is absorbed reasonably well.

An indirect effect of the rolled surface source is its contribution of hydrogen to remelted metal through recycled in-house scrap.

1.4 CONTROL OF ARTIFACTS ASSOCIATED WITH HYDROGEN

1.4.1 INTERDENDRITIC POROSITY

The characteristic defect in castings and ingots is interdendritic porosity, produced by hydrogen rejected from solution during solidification and trapped in the solidifying structure. In castings it reduces the mechanical properties and allows leaking. In ingots it is a source of blisters that develop in subsequent wrought products. The first essential in control is therefore to ensure that the metal is free from interdendritic porosity. This is accomplished by reducing the hydrogen content below certain critical values that depend on alloy composition and casting parameters. These critical or *threshold* values are below the typical hydrogen contents of metal prepared for casting and they must be reduced by degassing the liquid metal. The metal is degassed by purging it with an insoluble gas, bubbled through it to create a clean internal surface across which the dissolved hydrogen is evolved continuously and is swept away. The usual purge gas is nitrogen or argon with a small proportion of chlorine to flux the bubble surfaces thereby eliminating impedance by films formed by impurities in the nitrogen. Although degassing is effective when properly applied, a mathematical model of the process shows that the purge gas is used inefficiently.

Modern practice is to use *in-line* degassing, applying a purge gas countercurrent to the metal flow in customised vessels inserted in the transfer system[18–20] by which the metal is delivered from a furnace to ingot moulds.

This has advantages including more effective use of purge gas, better use of furnace capacity, less opportunity for the metal to reabsorb hydrogen and reduction of fume emission.

1.4.2 SECONDARY POROSITY

Ingots that are free from interdendritic porosity are still subject to porosity from the spontaneous breakdown of hydrogen solutions, yielding micropores.[12] It is convenient to refer to porosity from this origin as *secondary porosity* to distinguish it from interdendritic porosity. Secondary porosity persists through thermal and mechanical treatments and it can be inflated and reorganised into macroscopic defects at any stage of production where opportunity permits. This is particularly associated with prolonged heating at high temperatures to homogenise the metal and with heat-treatments in which the metal can absorb large quantities of hydrogen, such as solution treatment of age-hardening alloys containing magnesium in air furnaces polluted with sulphur dioxide.

1.4.3 EXTRANEOUS SITES FOR DEFECTS

Adventitious inclusions and faults introduced in the course of manufacture can become further sites for defects related to hydrogen.

Oxide Inclusions

Oxide from remelt metal and surface oxide on liquid metal incorporated in liquid metal delivered to ingots become strings of intermittent sealed discontinuities in thin flat rolled products and extrusions that can be inflated, forming blisters by hydrogen diffusing into them during intermediate or final heat-treatments. Comprehensive metal treatment systems have been developed from in-line degassing systems that can remove oxide fragments from liquid metal before casting.

Mechanical Faults

Examples of internal mechanically-generated defects stabilised by hydrogen from solution are cracks initiated by oversized intermetallic particles remaining from incomplete dissolution of master alloys, areas of poor adhesion at the interface of roll-bonded clad sheet, and double-skinning and back-end defects in extrusions. Control of such defects requires revision of fabrication practices.

1.4.4 INHERITED CHARACTERISTICS

To maintain the manufacture of aluminium products free from defects, production routes must be conceived as a whole because finished products inherit characteristics imparted to the metal at a different stage in a production sequence. One example is inflation of blisters during the heat-treatment of wrought metal in response to porosity in the ingot precursor. An example in reverse is the return of in-house scrap from rolled products to remelting furnaces, introducing a source of hydrogen from the contamination on its surface.

1.5 REFERENCES

1. D. E. J. Talbot, *International Met. Reviews*, 1975, **20**, 166.
2. F. D. Rossini, D. D. Wagman, W. H. Evans, S. Levine and I. Jaffee, *Nat. Bur Standards,*

Selected Values of Chemical Thermodynamic Properties, 1952.

3. W. Eichenauer and A.Pebler, Z. *Metallkunde*, 1957, **48**, 373.
4. D. E. J. Talbot and P. N. Anyalebechi: *Mat. Sci. and Tech.*, 1988, **4**, 1.
5. M. J. Stott and E. Zaremba, *Phys. Rev. B*, 1980, **22**, 1564.
6. C. E. Ransley and D. E. J. Talbot, J. *Ins. Metals*, 1955-56, **84**, 445.
7. C. E. Ransley, D. E. J. Talbot and H. Barlow: *J. Ins. Metals,* 1957-58, **86**, 212.
8. W. R. Opie and N. J. Grant, *Trans. AIME,* 1950, **188**, 1237.
9. P. N. Anyalebechi, D. E. J. Talbot and D. A. Granger, *Met. Trans.,* 1989, **20B**, 523.
10. C. G. McCracken, *'The Intrinsic and Extrinsic Solubility of Hydrogen in Aluminium-Lithium Based Alloys'*, Ph. D. Thesis, Brunel University, 1994.
11. C. G. McCracken, D. E. J. Talbot and J. Skov Pedersen, International Conference, Chicago 1992, TMS 1993, 481-87.
12. D. E. J. Talbot and D. A. Granger, *J. Inst. Metals*, 1963-64, **92**, 290.
13. J. Crank, *The Mathematics of Diffusion*, Oxford, 1956 (Clarendon Press).
14. O. Kubaschewski and C. B. Alcock, *Metallurgical Thermochemistry,* New York, 1979, (Pergamon press).
15. K. Wefers and G. M. Bell, *Technical Paper No. 19,* Alcoa Technical Center, Pa 15069, USA, 1972.
16. D. E. J. Talbot and J. D. R. Talbot, *'Corrosion Science and Technology'*, Boca Raton, Fa: 1997 (CRC Press).
17. P. T. Stroup, US Patents Nos 2092033 and 2092034.
18. L. C. Blayden and K. J. Brondyke, *Light Metals*, AIME, 1973, 493.
19. M. V. Brant, D. C. Bone and E. F. Emley, *J. Met.*, March, 1971, 48.
20. R. E. Miller, L. C. Blayden, M. J. Bruno and C. E. Brooks, *Light Metals*, AIME., 1978, **2**, 481.

2. Aluminium and its Alloys

Aluminium is a relatively expensive metal, because its extraction from the mineral, bauxite, is energy-intensive. The metal price fluctuates in response to supply and demand expressed in the metal exchanges through which it is traded. Prices of commercial grades of the metal and of alloys also vary according to composition and form, because they include the costs of alloying components, casting and fabrication.

The literature concerned with the structures, properties and applications of aluminium and its alloys is vast and beyond the scope of the present text. Comprehensive information in concise user friendly form is given in specialised texts,[1-3] especially those published by The American Society for Metals[1] and The Aluminium Association.[2] The following brief description summarises features of aluminium and some representative alloys based on it relevant to the present context.

2.1 STRUCTURE AND PROPERTIES OF ALUMINIUM

2.1.1 PHYSICAL STATES

The relative abundance of the stable isotope, ^{27}Al in the natural element is virtually 100% and there are no allotropes. The melting point is 933.4 K (660.4°C)[3] and the boiling point is estimated as 2767 K (2494°C).[4] The vapour pressure is negligible at normal industrial temperatures, e.g. it is $< 7.4 \times 10^{-9}$ kPa (7.4×10^{-11} atm) at temperatures < 1000 K (727°C).[4] The heat of fusion is 10.71 ± 0.21 kJ mol^{-1} and the heat capacity of the liquid, C_p, is 31.76 J mol^{-1} K^{-1} and is insensitive to temperature.[3]

The density of solid pure aluminium, calculated from the relative atomic mass and the lattice parameter is 2699 kg/m^3, in good agreement with measured values of 2698 ± 2 kg/m^3 for polycrystalline aluminium.[4] The density, ρ, of commercial grades is given by:

$\rho = [2698 + 2 \times 10^4 \text{(mass fraction of iron)}]$ kg/m^3

The density of very pure liquid aluminium (99.96%) is 2368 kg/m^3 at the melting point,[5] 933 K (660°C), diminishing approximately linearly to 2304 kg/m^3 at 1173 K (900°C).

Table 2.1 Equilibrium Fractions of Thermal Vacancies in Pure Aluminium, n_v mol^{-1}

Temperature	T/K	473	573	673	773	873	933
	°C	200	300	400	500	600	700
n_v mol^{-1}		8×10^{-8}	6×10^{-6}	4×10^{-7}	2×10^{-4}	7×10^{-4}	9×10^{-4}

Source: (Simmons and Balluffi[6])

2.1.2 CRYSTAL STRUCTURE

Aluminium crystallises in the face centred cubic structure, in which there are four atoms, four octahedral interstices and eight tetrahedral interstices per unit cube. The lattice parameter[4] is 0.40496 nm^3 and simple geometry yields 0.286 nm for the atomic diameter, considered as the closest distance between the centres of atoms, and 0.0592 nm and 0.0303 nm respectively for the inscribed radii of the octahedral and tetrahedral intersticies. Measured values for the equilibrium population of thermal vacancies are given by Simmons and Balluffi[6] and by Altenpohl.[7] The maximum vacancy population is 9×10^{-4} at the melting point and it diminishes rapidly with temperature, as given in Table 2.1, a significant factor in the generation of hydrogen porosity, as explained in Chapter 9, Section 9.3.1.2.

2.1.3 CHEMICAL PROPERTIES

Pauling[8] assigned aluminium the value 1.5 in his scale ranging from 0.7 to 4.0 for the most electropositive and most electronegative elements respectively, so that although elemental aluminium is clearly metallic, it is on the borderline between ionic and covalent in character in its compounds. Aluminium exists only in the oxidation state Al(III) in its solid compounds and in aqueous solution and in this state it is one of the most stable chemical entities known, as illustrated by the high value of the Gibbs free energy of formation for the oxide:[9]

$$2Al + 1\frac{1}{2}O_2 = Al_2^{III}O_3$$

$$\Delta G^{\ominus} = -1117993 - 10.96\,T\,\log T + 244.5T\ \mathrm{J} \qquad (2.1)$$

and of the standard electrode potentials, $E^{\ominus}$, for formation of ions in aqueous solutions:[10]

$$Al = Al^{3+} + 3e^- \qquad E^{\ominus} = -1.64\ \mathrm{V\ (SHE)} \qquad (2.2)$$

$$Al + 4OH^- = AlO_2^- + 2H_2O + 3e^- \qquad E^{\ominus} = -2.35\ \mathrm{V\ (SHE)} \qquad (2.3)$$

These values show that aluminium is intrinsically more reactive than any other common engineering metal except magnesium and its relative permanence is attributable to the protective oxide films that form upon it. Some of the more significant aluminium

Table 2.2 Phase Relationships for Some Binary Aluminium Alloys

System	Maximum Solid Solubility Wt.% Alloy Element	Eutectic Composition Wt.% Alloy Element	Eutectic Temperature °C	Eutectic Components
Aluminium–Copper	5.7	33.2	548	$Al^* + CuAl_2$
Aluminium–iron	< 0.1	1.7	655	$Al^* + FeAl_3$
Aluminium–Lithium	5.2	9.9	600	$Al^* + LiAl$
Aluminium–Magnesium	14	35	450	$Al^* + Mg_2Al_3$
Aluminium–Manganese	1.8	1.9	660	$Al^* + MnAl_6$
Aluminium–Silicon	1.65	12.5	580	$Al^* + Si^\dagger$
Aluminium–Zinc	82.8	94.9	382	$Al^* + Zn^\ddagger$

* Containing maximum solubility of second element in solid solution.
† Containing 0.5 wt.% of aluminium in solid solution.
‡ Containing 1.1 wt.% of aluminium in solid solution.

Source: (Aluminium: Properties and Physical Metallurgy[1])

compounds in metallurgy are the oxides and hydroxides, that are described in detail in Chapter 8, Section 8.3.2.3, in relation to the absorption of hydrogen by aluminium from humid atmospheres.

2.2 SUMMARY OF COMMERCIAL ALLOYS AND APPLICATIONS

Pure aluminium is deficient in two respects, mechanical strength and elastic modulus and aluminium alloy development was driven by the need to improve them without sacrificing other qualities, e.g. to improve strength for aerospace, marine and civil engineering applications without losing corrosion resistance. The success of these endeavours has secured the status of aluminium alloys as second only to steels in economic value.

2.2.1 ALLOY SYSTEMS

Copper, lithium, magnesium and zinc have fairly extensive solubilities in solid aluminium but the solubility of many other elements is very limited. Most binary systems with aluminium are characterised by limited solubility of the second metal and a eutectic in which the components are the saturated solution and an intermetallic compound, Al_xM_y.

Aluminium is prolific in the number and variety of intermetallic compounds it can form because of its moderately high electronegativity and high valency. Critical features of some common binary systems are summarised in Table 2.2.

Commercial alloys are based on multicomponent systems that reflect characteristics of the component binary systems but also exhibit features due to interaction between the alloying elements, especially in introducing additional intermetallic compounds and in modifying solubilities.

In alloy formulation, strength can be imparted by:

1. Reinforcement with intermetallic compounds.
2. Solid solution strengthening.
3. Work hardening.
4. Precipitation hardening (ageing).

There are more than a hundred current wrought and cast alloy compositions but they belong to a relatively few series with particular characteristics. A representative selection is given in Table 2.3, using the internationally recognized *AA* (Aluminium Association) designations. The following brief descriptions can be supplemented by the more detailed treatment given by Polmear.[11]

2.2.2 ALLOYS USED WITHOUT HEAT–TREATMENT

2.2.2.1 Commercial Pure Aluminium Grades (AA 1100 Alloy Series)

Commercial grades of pure aluminium are actually dilute alloys. The iron and silicon contents of the aluminium product from the Hall-Herault process when it was first introduced raised the properties of the metal to values suitable for a wide variety of general applications, including domestic and catering utensils, packaging foil, some chemical equipment and architectural applications. Although modern practice produces higher purity metal, grades of aluminium with similar iron and silicon contents are still offered as general purpose materials. Conventionally, these alloys are designated by specifying the aluminium content, e.g. the most common alloy, AA 1100 in Table 2.3 has a minimum aluminium content of 99.0% and typical iron and silicon contents of 0.45 and 0.25% respectively. The metal is strengthened by dispersed intermetallic compounds, notably Fe_3SiAl_{12}, $FeAl_3$ and the metastable $FeAl_6$, raising the 0.2% offset yield and tensile strengths to the values given in Table 2.4. Additional strength can be and usually is imparted by work hardening as illustrated by the values for 75% cold reduction in thickness, given in the same table.

2.2.2.2 The Aluminium-Manganese Alloy AA 3003

The alloy, AA 3003, has applications that overlap those of AA 1100 but it has greater strength as indicated in Table 2.4. It is produced by adding 1.2% of manganese to the AA 1100 composition. The intermetallic compounds are modified to $(Fe, Mn)_3SiAl_{12}$ and $(Mn, Fe)Al_6$. Compounds containing manganese are also present as a fine dispersoid distributed throughout the aluminium matrix.

Table 2.3 Typical Chemical Compositions of Representative Aluminium Alloys

Alloy	Alloy Elements Mass%							Examples of Applications*
	Silicon	Copper	Manganese	Magnesium	Zinc	Lithium	Others	
AA 1050	Minimum of 99.50% Aluminium – Principal Impurities Iron and Silicon							Architectural Applications
AA 1100	Minimum of 99.00% Aluminium – Principal Impurities Iron and Silicon							Cooking Utensils, Foil
AA 2024	-	4.4	0.6	1.5	-	-		Airframe Structures
AA 3003	-	0.12	1.2	-	-	-		General Purpose Alloy, Foil
AA 3004	-	-	1.2	1.0	-	-		Beverage Cans
AA 5005	-	-	-	0.8	-	-		Architectural Applications
AA 5050	-	-	-	1.4	-	-		General Purpose Alloy
AA 5182	-	-	0.35	4.5	-	-		Beverage can Ends
AA 5456	-	-	0.8	5.1	-	-	0.12 Cr	Transportation, Structures
AA 6061	0.6	0.28	-	1.0	-	-	0.20 Cr	Extrusions, Beer Containers
AA 7075	-	1.6	-	2.5	5.6	-	0.23 Cr	Airframe Structures
AA 7072	-	-	-	-	1.0	-		Cladding to Protect 7075 Alloy
AA 2090	-	2.6	-	0.01	-	2.1	0.12 Zr	Aerospace Applications
AA 2091	-	2.1	-	1.5	-	2.1	0.12 Zr	Aerospace Applications
AA 8090	-	1.1	-	0.65	-	2.3	0.12 Zr	Aerospace Applications
AA 2195	0.14	4.75	-	0.4	-	1.25	0.03 Zr + 0.4 Ag	Aerospace Applications
AA 319	6.0	3.5	-	-	-	-		General Purpose Castings
AA 380	8.5	3.5	-	-	-	-		Die Castings
AA 356	7.0	-	-	0.35	-	-		Age-Hardenable Castings
AA 390	17	4.5	-	0.55	-	-		Automobile Cylinder Blocks

* Selected applications. Many alloys are multi-purpose.

Table 2.4 Typical Strengths of Some Aluminium Alloys at 25°C

Alloy	Condition	0.2% Yield Strength MPa	Tensile Strength MPa
99.99%	Annealed (Softened by Heating)	10	45
AA 1100	Annealed	35	90
	75% Cold Reduction (by Rolling)	150	165
AA 3003	Annealed	40	110
	75% Cold Reduction	185	200
AA 3004	Annealed	70	180
	75% Cold Reduction	250	285
AA 5005	Annealed	40	125
	75% Cold Reduction	185	200
AA 5050	Annealed	55	145
	75% Cold Reduction	200	220
AA 5182	Annealed	130	275
	75% Cold Reduction	395	420
AA 5456	Annealed	160	310
	Fully Strain Hardened†	255	350
AA 6061	Annealed	55	125
	Solutionized at 532°C. Aged 18 hrs. at 160°C	275	310
AA 2024	Annealed	75	185
	Solutionized at 493°C. Aged 12 hrs. at 191°C	450	485
AA 7075	Annealed	105	230
	Solutionized at 482°C. Aged 24 hrs. at 121°C	505	570
†Using Special Temper Procedure to Minimize Structural Instability			

2.2.2.3 Aluminium–Magnesium Alloys (AA 5000 Series)

Magnesium both imparts solid solution strengthening and enhances the ability of the metal to work harden, illustrated for alloys AA 5005, 5050, 5182 and 5456 in Table 2.4. They have practical advantages in their good formability, high rate of work hardening, weldability and corrosion resistance. A reservation on the magnesium content is set by the instability of the solid solution with respect to precipitation of Mg_2Al_3 that can be significant for magnesium contents greater than about 3%, during long storage at ambient temperatures; the precipitation is at grain boundaries and can render the metal susceptible to stress-corrosion cracking and to intergranular corrosion. Unfortunately, the instability is increased by work hardening that is one of the alloys most useful attributes but it can be ameliorated with some loss of strength by special tempering procedures.

2.2.2.4 The Aluminium–Magnesium–Manganese Alloy AA 3004

The alloy AA 3004 is strengthened by both magnesium and manganese and has an optimum combination of formability, and work hardening for application as beverage cans that are deep drawn and must have sufficient strength in very thin gauges to withstand forces imposed on them by filling and by internal gas pressure.

2.2.3 HEAT TREATABLE (AGEING) ALLOYS

The strongest aluminium alloys are strengthened by precipitation hardening. The principle exploits the diminishing solubility of certain solutes with falling temperature. The strength is developed by controlled decomposition of supersaturated solid solutions. An alloy is heated to and held at a high temperature to allow the solutes to dissolve, an operation called *solution treatment*, and then rapidly cooled to retain them in supersaturated solution. Rapid cooling is usually accomplished by immersing the hot metal in cold water, i. e. *quenching*. The metal is subsequently reheated to a constant moderate temperature and the unstable supersaturated solution rejects excess solute in the form of finely dispersed metastable precursors of the final stable precipitate. These precursors are of suitable forms, sizes and distributions to impede the movement of dislocations in the lattice that effect plastic flow. The process occupies a considerable time and is called *ageing*. The improvement in mechanical properties is illustrated by the values given in Table 2.4 for annealed and age-hardened versions of alloys AA 6061, A 2024 and AA 7075.

Not all aluminium alloy systems are amenable to strengthening in this way. Standard commercial alloys are based on the aluminium-copper-magnesium, aluminium-zinc-magnesium and aluminium-magnesium silicon systems. In these multicomponent systems, the ageing process depends on the decomposition of supersaturated solutions of more than one solute from which complex intermetallic compounds are precipitated. A typical sequence of events during ageing is:

1. Assembly of solute atoms in groups, called *Guinier Preston* (GP) zones, dispersed throughout the matrix at spacings of the order of 100 nm.
2. Loss of continuity between the zones and the metal lattice in some but not all crystallographic directions, yielding transition precipitates usually designated by primed letters, sometimes with their formulae in parenthesis.
3. Complete loss of continuity with the matrix, yielding particles of stable precipitates. The sequence is conveniently described by an equation of the general form:

Solid Solution $\rightarrow$ GP $\rightarrow$ θ' $\rightarrow$ precipitate (2.4)

The detail varies widely from system to system and is more complicated than eqn 2.4 indicates, depending on metal composition, precipitate morphology and ageing temperature.

The metal becomes stronger as dislocation movement is inhibited by matrix lattice strains building up around the GP zones and transition precipitates but as these transition species are progressively replaced by equilibrium precipitate, the lattice strains are relaxed and the metal softens. There is thus an optimum ageing period for maximum strength, i.e. *peak hardness*. Metal that has been aged for less than this period is said to be *underaged* and metal that has passed the peak is *overaged*. These conditions have important implications in corrosion and stress-corrosion cracking.

The formation of transition species and their replacement by equilibrium precipitates are both accelerated at higher temperatures. The consequence is that as the ageing temperature is raised, the peak hardness is reached in progressive shorter periods of time but its value diminishes. There is thus an optimum thermal cycle to secure good results economically.

2.2.3.1 Aluminium–Copper–Magnesium Alloys (e.g. AA 2024)

The ageing sequence is usually represented by:

$$\text{Solid Solution} \rightarrow \text{GP} \rightarrow S'(\text{Al}_2\text{CuMg}) \rightarrow S(\text{Al}_2\text{CuMg}) \qquad (2.5)$$

where $S'(\text{Al}_2\text{CuMg})$ and $S(\text{Al}_2\text{CuMg})$ are transition and stable precipitates respectively.

The classic examples of alloys in this system are AA 2024 and its variants that are staple materials for airframe construction. The solution treatment temperature is restricted to 493°C to avoid exceeding the solidus temperature and ageing is at 191°C for a period of between 8 and 16 hours to suit the form of the material.

2.2.3.2 Aluminium–Zinc–Magnesium Alloys (e.g. AA 7075)

The alloys based on this system are some of the strongest produced commercially and are also staple materials for airframe construction; the characteristic alloy is AA 7075. Particular compositions, solution treatments and ageing programs have been developed to secure practical benefits of strength, consistency of properties and minimum susceptibility to stress-corrosion cracking. Detailed assessment of the ageing sequences is difficult because the transition and final species encountered depend on composition, initial microstructure, speed of quenching and ageing temperature. Schemes suggested include:

$$\text{Solid Solution} \rightarrow \text{GP} \rightarrow \eta' \rightarrow \eta(\text{MgZn}_2) \text{ or } T(\text{Mg}_3\text{Zn}_3\text{Al}_2) \qquad (2.6)$$

The heat-treatment is critical. Alloy AA 7075 is typically solution treated at 482°C, just below the solidus temperature, and aged for 24 hours at 121°C.

2.2.3.3 Aluminium–Magnesium–Silicon Alloys (e.g. AA 6061)

The precipitating phase in this system is Mg_2Si and the ageing scheme is:

$$\text{Solid Solution} \rightarrow \text{GP} \rightarrow \beta'(\text{Mg}_2\text{Si}) \rightarrow \beta(\text{Mg}_2\text{Si}) \qquad (2.7)$$

Alloys in this system are not as strong as those based on the aluminium-copper-magnesium and aluminium-magnesium-zinc systems but they have the following compensating advantages:

1. They are easy to hot extrude into sections.
2. In the soft condition they are ductile and accept deep drawing into hollow shapes.
3. They are the most corrosion resistant of the age-hardening alloys.

These attributes recommend them for uses including architectural fittings such as window frames, food handling equipment, beer barrels and transport applications. The most widely applied alloy is AA 6061 but there are other related formulations. AA 6061 is solution treated at 532°C and aged for 18 hours at 160°C.

2.2.3.4 Aluminium Alloys Containing Lithium

In recent years, a range of multicomponent alloys has been developed for aerospace applications based on alloys containing lithium. Some examples are given in Table 2.3. One of the objectives is to match the properties of AA 2024 and AA 7075 alloys but with lower density and higher modulus, both of which are promoted by the use of lithium. The basic ageing sequence:

$$\text{Solid Solution} \rightarrow \text{GP} \rightarrow \delta'(\text{Al}_3\text{Li}) \rightarrow \delta(\text{AlLi}) \tag{2.8}$$

is complicated by the presence of other alloy components needed to secure the required properties, notably copper and magnesium, that lead to schemes that include additional hardening phases, $T'(\text{Al}_2\text{CuLi})$ and $S'(\text{Al}_2\text{CuMg})$.

These alloys are expensive and justified only where mass saving is a critical economic factor as in aerospace applications. Future interest in them depends on assessment of their long term integrity but they are available for exploitation.

2.2.4 CASTING ALLOYS

The most common casting alloys are based on the eutectic aluminium-silicon system and are characterized by fluidity of the liquid metal and low contraction on solidification. There are many composition variants and Table 2.3 lists alloys AA 319, 380 and 356 as examples in widespread use; the function of the copper content in AA 319 and AA 380 is to improve strength and machinability at some sacrifice of corrosion resistance. The magnesium content in alloy AA 356 offers the facility of precipitation hardening by Mg_2Si. Alloy AA 390 is a customized special alloy for automobile cylinder blocks.

2.3 REFERENCES

1. John E. Hatch, ed., *Aluminium: Properties and Physical Metallurgy*, American Society for Materials, Metals Park, Ohio.
2. *Aluminium Standards and Data*, The Aluminium Association, Washington, D.C.
3. L. F. Mondolfo, *Aluminium Alloys: Structure and Properties*, Butterworths, Boston, 1976.

4. D. R. Stull and H. Prophet, JANAF *Thermochemical Tables*, US Department of Commerce, Washington, D.C.
5. E. Gebhardt, M. Becker and S. Dorner, *Z. Metallkunde*, **44**, 1953, 573.
6. R. O. Simmons and R. W. Balluffi, *Phys. Rev.*, **117**, 1960, 2.
7. D. Altebpohl, *Aluminium*, **37**, 1961, 401.
8. L. Pauling, *The Nature of the Chemical Bond*, Cornell University Press, Ithica, 1967.
9. O. Kubaschewski, E. Ll. Evans and C. B. Alcock, *Metallurgical Thermochemistry*, Pergamon Press, Oxford, 1967.
10. L. L. Shrier, *Corrosion*, **21**, 20 Butterworths, London, 1963.
11. I. J. Polmear, *Light Alloys*, 3rd Edition, Edward Arnold, London, 1995.

3. Summary of Manufacturing Routes and Procedures

The aluminium industry includes all stages in the extraction of the metal and its conversion to a wide range of semi-finished products, comprising the alloy groups introduced in Chapter 2 supplied in various forms including sheet, plate, foil, extruded sections, castings and forgings. This brief summary of processes by which they are manufactured forms the context in which to consider the behaviour of hydrogen in the metal. It covers essential principles of extraction, casting and fabrication of the metal and its alloys into standard forms.

3.1 EXTRACTIVE METALLURGY OF ALUMINIUM

The naturally occurring sources of aluminium are bauxites of various kinds that are formed by the weathering of aluminium-bearing rocks, leaving mainly alumina hydrates with characteristic impurities, mainly iron oxides, titania and silica.

The Gibbs standard free energy change for the reduction of alumina to aluminium is so negative that electrolytic reduction of alumina dissolved in a non-aqueous electrolyte is the only viable option. The process in universal use is the Hall-Herault process in which the electrolyte is cryolite, Na_3AlF_6. The alumina must be pure because impurities entering the metal by selective electrolysis cannot be removed economically.

3.1.1 EXTRACTION OF ALUMINA FROM BAUXITE BY THE BAYER PROCESS[1]

Crushed bauxite is digested with sodium hydroxide solution at 140–180°C under a steam pressure of about 1.5 MPa (15 atm.) for several hours. The alumina is extracted as soluble sodium aluminate, leaving an insoluble residue containing most of the impurities. The clear filtered sodium aluminate solution is diluted and cooled to decompose it and seeded to nucleate alumina trihydrate. The trihydrate is filtered off and calcined to the anhydrous alumina, corundum.

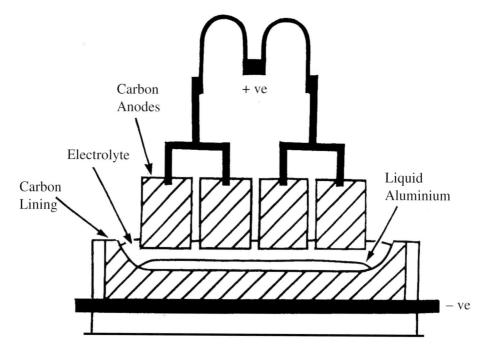

Fig. 3.1 Schematic diagram of Hall-Herault cell for electrolytic extraction of aluminium from pure alumina dissolved in cryolite at 950°C.

3.1.2 Extraction of Aluminium by the Hall-Herault Process[1]

3.1.2.1 The Reduction Cell

Aluminium is deposited by electrolysis of calcined alumina from the Bayer process dissolved in liquid cryolite at 950°C in a cell illustrated schematically in Figure 3.1. It is a thermally insulated steel box lined with carbon blocks holding a bath of liquid cryolite about 30 cm deep. The primary cathode reaction is:

$$Al^{3+} + 3e^- = Al(liquid) \tag{3.1}$$

The cathodic product is liquid aluminium and because it is denser (2.3 g cm^{-3}) than the electrolyte (2.1 g cm^{-3}) it is deposited as a layer on the bottom of the cell, forming the effective cathode. It is tapped periodically by suction through a pipe inserted through the electrolyte.

The anode comprises carbon blocks dipping into the cryolite from above. The cryolite is covered by a frozen crust that is broken at intervals to replenish alumina as it is depleted. The anodic reaction is uncertain but the species discharged is oxygen and it is represented by:

$$4AlO_3^{3-} = 2Al_2O_3 + 3O_2 + 12e^- \tag{3.2}$$

The carbon of the anode depolarises the anode reaction:

$$C + O_2 = CO_2 \tag{3.3}$$

so that the carbon anode blocks are consumed and must be replaced.

The cell size is limited by the heat liberated in the electrolyte and a typical cell operates at 4.5 V and 100000 amps. Cells are connected in series called *pot lines* with total potential differences suitable for an economic power supply. The operating potential of a cell is the sum of the equilibrium potential, 1.2 V, and potential differences due to polarisation and ohmic resistance so that the potential efficiency is about 30%. The Faradaic equivalent of the operating current is 0.8 tonnes daily but side reactions reduce current efficiency to 85%.

Process costs are dominated by the cost of the electrical energy to satisfy the high valency of aluminium that requires 3 moles of electrons, i.e. 289500 coulombs, for every mole of aluminium produced and to compensate for the energy inefficiencies. These costs preclude production of aluminium using electricity at normal commercial rates. Hence the smelters are usually located at plentiful sources of electricity too remote for transmission to large centres of population. Most use hydro electric power or electricity generated from oilfield gas.

3.1.2.2 Metal Purity

The standard electrode potentials for decomposition of most other metal oxides are much less negative than that for alumina and they are preferentially reduced even if their activities in the electrolyte are low. Thus impurity oxides entering the electrolyte from the alumina feedstock, consumable carbon anodes or other cell materials appear as corresponding metal impurities in the product. Hence the feedstock and the materials used to construct the cell must be of very high purity because subsequent purification of the aluminium product is not commercially viable. A small quantity of sodium is derived from the cryolite, < 0.01% but it is easy to remove subsequently by selective oxidation or chlorine treatment. Despite these considerations modern plants can yield metal with a consistent purity of up to 99.99%.

3.2 INGOT CASTING

The direct chill (DC) semicontinuous process is the dominant method for casting ingots for fabrication. Its essential feature is efficient cooling of ingots by direct impingement of a water spray that can yield sound ingots with acceptable structures. It is operated within an integrated system that also includes melting, preparation and delivery of liquid metal.

3.2.1 MELTING AND PREPARATION OF THE LIQUID METAL

Liquid metal is usually prepared for casting in oil or gas fired reverberatory furnaces operated in tandem. One is a melting furnace with a capacity in the range, 20 to 40 tonnes, in which liquid metal from reduction facilities is received, solid metal charges are melted and master alloys are added. The melting furnace delivers metal by a short trough to one or two holding furnaces for final preparation. This entails making trimming additions, allowing large inclusions to settle out and adjusting the temperature to a value with a superheat appropriate for the alloy, typically 680–700°C.

3.2.2 DELIVERY OF LIQUID METAL FROM FURNACE TO MOULDS

The liquid metal is transferred from the holding furnace and distributed to the ingot moulds by refractory lined launders with wide cross-sections, typically 20×20 cm, with subsurface entry from the furnace and subsurface exit to the moulds through a tube controlled by a submerged float valve. This maintains a quiescent liquid surface in the launder and eliminates free fall between different levels to avoid generating and entraining oxide. Metal is delivered from the same furnace to several moulds simultaneously.

Metal treatment devices are inserted in the transfer system, including filters to remove inclusions and in-line degassing systems.

The filters are of two kinds:

1. Consumable strainers made from woven glass fibre, ceramic foam plates or fritted refractory particles that are inserted in the troughs.
2. Permanent deep bed filters[2] that are heated vessels 50–70 cm deep packed with tabular alumina[2] flake supported on alumina balls to which small inclusions adhere by Van der Waals forces. Baffles direct the metal upwards through the filter bed.

In-line degassing systems were developed from deep bed filters and are described later, after the objectives and principles of degassing have been determined in Chapters 9 and 10.

3.2.3 THE DIRECT CHILL (DC) SEMICONTINUOUS PROCESS

3.2.3.1 Regular DC Casting

The essential features of the process are illustrated schematically in Figure 3.2. The mould is a shallow, water-cooled aluminium box, open at the top and initially closed at the bottom by a close-fitting aluminium starter block that can be retracted into a deep pit to withdraw an ingot as it solidifies. A copious flow of cooling water is supplied to the water jacket and discharged as a uniform spray directed towards the emerging ingot. Liquid metal is admitted from a launder to the mould by subsurface entry through a vertical refractory tube controlled by a float valve.

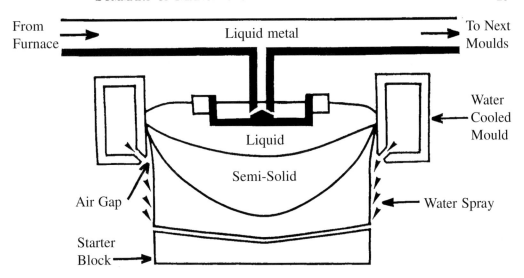

Fig. 3.2 Schematic diagram of direct chill (DC) semicontinuous casting process.

To start casting, the starter block is raised to close the open end of the mould and liquid metal is admitted. The first solid grows on the mould walls and the starter block, forming a shell around and below the liquid metal that seals the bottom of the mould. It is lowered on the starter block into the water spray which impinges directly on its surface, starting the solidification of an ingot that is withdrawn at a constant speed, supported on the starter block. The solid metal withdrawn from the mould is automatically replenished by liquid metal admitted through the float valve. Casting continues until the bottom block reaches the end of its travel. The rate at which an ingot is withdrawn is called the casting speed. Pure aluminium and some alloys can be cast at speeds in the range 40 to 100 mm per minute but crack-sensitive alloys must be cast at lower speeds. Moulds are interchangeable to match the ingot sections required, e.g. large rectangular sections, typically 500 × 1350 mm for rolling to plate or sheet and smaller circular sections, 180 mm in diameter, for extrusion.

The shell contracts as it descends in the mould, forming an insulating air gap so that it is reheated by sensible heat conducted from the interior. Liquid enriched in alloy components by selective freezing infiltrates the reheated shell and solidifies in the air gap, yielding exudations and inverse segregation.

3.2.3.2 Low Head Casting
The surface quality of ingots produced by regular direct chill casting is not always acceptable. It can be improved by restricting the time during which the shell is reheated.

This is accomplished by lowering the liquid metal delivery system so that the shell forms lower in the mould but it is difficult to start casting without bleed-out. One solution is to start in the regular way and then raise the mould[3] and another is to dissolve carbon dioxide in the cooling water to reduce the initial quench.[4]

3.2.3.3 Level Transfer Casting

In level transfer casting, the mould is raised and integrated with the launder system so that the liquid level is the same in both. The advantage of quiescent transfer from launder to mould is offset by lower surface quality because the mould length is fixed. It can be alleviated by a proprietary method in which a gas is introduced at the top of the mould to displace the start-of freeze location, reducing the effective mould length.[4]

3.3 ROLLING FLAT PRODUCTS

Sheet and plate are produced by rolling. The initial reduction is by hot rolling to minimise the power required and to develop favourable microstructures. The finishing stages are by cold rolling to control the gauge, produce good surface finish and develop strength by work hardening if required.

3.3.1 HOT ROLLING

The major faces of the ingots are milled to remove surface exudations (scalping) and cut into lengths to suit the hot mill facilities (rolling blocks). They are prepared for rolling by homogenisation, a thermal treatment to disperse heterogenieties in the microstructure of the as-cast metal to facilitate fabrication by improving its ductility. The thermal cycle is alloy specific with the maximum temperature ranging from 450 to 590°C depending on solidus temperatures. The rate of heating is determined by the available heat input and the thermal capacity of the rolling blocks. If the alloy is to be roll-bond clad with pure aluminium or an aluminium–zinc alloy for corrosion protection, prepared cover plates of the cladding metal, each typically 5% of the thickness of the composite are placed beneath and on top of the scalped ingot and held with steel straps to keep them in place in the preheating furnace. The straps are discarded when the composite is presented to the hot mill.

The preheated ingots are reduced to the required thickness in multiple passes. A typical facility comprises a reversing mill for the initial passes followed by a continuous mill for reduction to the final hot line gauge. The metal is lubricated by an emulsion of soluble oil in water directed into the roll gap and recirculated through a large reservoir. The lubricant is replaced at intervals of several weeks because it degrades forming metal soaps.

Metal for subsequent cold-rolling is reduced to a gauge suitable for coiling and wound into coils for onward handling. Plate products are finished at the prescribed thickness, cut to length and straightened by stretching by about 2% between hydraulically operated grips.

3.3.2 Cold Rolling

In hot rolling, the metal is chilled on the roll faces to a finishing temperature at which it is partly work-hardened and it must be softened by annealing before presenting it to a cold mill. The coils are heated in a furnace with an even temperature distribution, to raise their temperature to a value required to soften the metal, typically in the range 350 to 400°C according to alloy composition, taking several hours due to the high thermal capacity of the charge.

The annealed coils are cold-rolled on a multi-stand tandem mill with a mineral oil lubricant and recoiled at the required gauge. Further annealing and re-rolling are manipulated to produce sheet in the required gauge and temper. If required in the solution-treated condition, age hardening alloy sheet is cut to size, and solution-treated in a molten nitrate salt bath at recommended temperatures and water-quenched. Table 2.4 gives some examples of solution-treatments for alloys AA 6061, 2024 and 7075.

3.3.3 Foil Rolling

Elastic deformation of the rolls imposes a limit to the gauge attainable in cold rolling by pressure in the roll gap and further reduction is obtained by applying tension to the stock on the outgoing side of the mill. This is the basis of foil rolling which is carried out in coil using the rewind mechanism to apply tension.

3.4 FABRICATION OF RODS TUBES AND SECTIONS

3.4.1 Extrusion

Extrusion is forcing metal through a die of the required section by hydraulic pressure applied by a ram to a hot cylindrical ingot or 'billet' in a closed container. The process is designed to produce long rods, sections and tubes. Some alloys, such as members of the AA 6000 series are particularly suited to extrusion because they are ductile when homogenised and the extrusions can be age-hardened subsequently. Billets for extrusion can be heated in gas fired furnaces or by induction, allowing for adiabatic heating due to the work done in the die. The metal flow pattern is determined by friction on the container walls and by a dead zone around the die, leading to characteristic extrusion defects, which include double skinning and back-end defect, a long axial void in the last part of the extrusion which is discarded. Extrusions of age-hardening alloys are usually too long to be accommodated in salt baths and are therefore solution-treated in electric air circulation furnaces. A less expensive but less reliable alternative sometimes used is to quench extrusions as they emerge hot from the die.

3.4.2 ROD AND SECTION ROLLING

An alternative to extrusion, especially for heavier sections is hot or cold rod and section rolling, using roll pairs with mating circumferential grooves of the required shape. The metal is reduced by passing it through successively smaller grooves.

3.4.3 TUBE AND WIRE DRAWING

Drawing is cold reduction of a long rod or section, usually circular by pulling it through dies with successively smaller apertures of similar shape. It is especially suited to the production of rods and tubes with close tolerances and good surface finish and where strength is developed by work hardening. Extruded or rolled rod and extruded tube shell are typical starting intermediates. The bores of tubes are supported by internal die cores.

3.5 FOUNDRY PRODUCTS

The foundry industry is dispersed among many facilities, some of which are departments within integrated engineering establishments where operations such as melting and metal treatment are conducted on a relatively small scale and without the metallurgical specialisms available to large international wrought aluminium manufacturers.

Foundry alloys are often produced from secondary metal supplied for remelting. The metal is melted in small furnaces together with recycled metal from runners and risers that are cut from finished castings. The scale of operations is seldom large enough to enable application of in-line metal treatments and the liquid metal is often accessible for treatment only in crucibles in which it is conveyed from the furnaces to the moulds.

Sand Casting
In sand casting, sand is rammed around a wooden or plaster pattern in a mould box that is split to allow the pattern to be removed. Provision is made for filling, feeding and venting the mould, by carving out suitable runners and risers. Various techniques are available to bind the sand, to mould cores to form internal features and to place chills to direct local solidification patterns. To reduce risks of the synergistic effects of hydrogen and solidification contraction, castings are designed to allow free access of the liquid to

all parts until solidification is complete. Before filling, sand moulds are thoroughly dried to eliminate residual moisture that is a source of hydrogen.

Die Casting

Die casting is the injection of metal into a permanent shaped closed die from which successive castings are ejected when solidified. It is suitable for mass producing identical small parts. There are two forms, gravity die casting in which the die is filled passively under gravity and pressure die casting in which the die is filled with metal shots under high pressure. The metal is supplied from a heated sump that is replenished as necessary.

3.6 FORGING

Forging is working a metal part to a predetermined shape by hammering, pressing or upsetting the hot metal, singly or in combination. Hammer forging is shaping by repeated blows under a forging hammer. Press forging is shaping by pressure applied in a forging press either in the open or in an impression die. Upset forging is manipulating the piece in such manner as to increase its cross section. Hot working may be followed by cold-coined forging in which the piece is restruck to meet dimensional tolerances and outlines or to increase hardness in non-heat-treatable alloys.

Forging stock is cut from direct chill cast ingots or rolled rod or bar with sections to suit the finished forgings. The ingots are cast subject to the same quality considerations as ingots for rolling or extrusion.

3.7 RECYCLING

Recycling is an important economic aspect of the aluminium industry to recover the value of the metal, to conserve the energy needed to smelt equivalent quantities of new metal and to eliminate associated carbon dioxide emissions. Sources of material for recycling include:

I.   In-house scrap, including trimmings from flat-rolled products, extrusion discards, runners and risers from castings.
2.   Residues of sheet from which blanks are punched.
3.   Returned beverage cans.
4.   Miscellaneous scrap collected from the field.
5.   Dross skimmed from liquid metal and millings.

In-house scrap, returned discards and beverage cans provide regular supplies of metal with known compositions that can be remelted as part of charges with similar specifications. The scrap usually has a high surface to volume ratio and to minimise losses by oxidation it is submerged in liquid metal so that it melts out of contact with air.

Recycling beverage cans is particularly economical because most of the expensive magnesium content is recovered from the AA 3004 and AA 5182 alloys from which they are produced.

Miscellaneous scrap collected from the field is of lower value because the composition is variable and uncertain and it is often corroded or contaminated. Specifications with generous ranges are available which can be met by appropriate additions to the compositions obtained on a batch by batch basis. A representative treatment for miscellaneous light scrap is to remove tramp iron magnetically, separate tramp copper, shred and steam clean it and melt it by submersion in a heel of liquid metal. It is treated by in-line devices similar to those described in Chapter 10 to remove hydrogen and contaminants such as sodium and cast into ingots of a size suitable for foundry use. Typical applications are general purpose castings.

Metal is recovered from dross for miscellaneous uncritical applications. It can be separated from the oxide as a cohesive layer by melting and treating it with chlorine under liquid halide fluxes. Millings produced in preparing cast ingots for rolling contain surface segregates rich in alloy components which can be recovered by remelting the material and adding it to subsequent furnace charges.

3.8 REFERENCES

1. T. G. Pearson, *"The Chemical Background of the Aluminium Industry"*, The Royal Society for Chemistry, London, 1955.
2. K. J. Brondyke and P. D. Hess, *Trans. AIME*, **230**, 1964, 1553.
3. D. Altenpohl, *Proceedings of 5th International Light Metals Conference*, Leoben, 1968, 367–373.
4. R.E. Spear and H. Yu, *Aluminium*, **60**, 1984, 440.

4. Hydrogen Occlusion in Aluminium and its Alloys

4.1 DILUTE SOLUTIONS OF HYDROGEN IN ALUMINIUM

Size factors and valency considerations restrict the accommodation for dissolved hydrogen to the lattice interstices in the solid or the corresponding less regular interatomic spaces in the liquid. The largest potential site for hydrogen in solid aluminium is the octahedral interstice, which has an inscribed radius of 0.059 nm calculated from the spacing of the atoms in the aluminium FCC lattice[1, 2] and, since the change in the volume of the metal on melting, is ~ 12%, the average radius of interatomic spaces in the liquid is ~ 4 % larger. Spaces with these radii can accommodate the neutral hydrogen atom, characterised by the Bohr radius, 0.053 nm,[3, 4] but not the molecule which comprises two atoms with an internuclear distance of 0.074 nm.[3, 4] Hence the gas is expected to dissociate on dissolution in both the solid and the liquid metal:

$$H_2 \text{ (gas)} = 2H(\text{solution in metal}) \tag{4.1}$$

4.1.1 THERMODYNAMICS OF DISSOLUTION

The equilibrium constant, K, for eqn 4.1 is:

$$K = \frac{(a_H)^2}{a_{H_2}}$$

Rearranging yields:

$$a_H = K^{1/2} \cdot (a_{H_2})^{1/2} \tag{4.2}$$

Application of standard theory to eqn 4.1 yields quantitative relations for the dependence of the equilibrium on pressure and on temperature, referring the activities of the diatomic gas and of the atomic solute to selected standard states.

4.1.1.1 Selection of Standard States

Standard State for Gaseous Hydrogen
The conventional standard state for a gaseous element[5-7] is the pure gas at a standard pressure, $p^{\ominus}$, usually selected as 101 kPa (1 atm.) and since hydrogen is virtually ideal at temperatures of interest, its activity, a_{H_2}, is proportional to its pressure, p, so that:

$$a_{H_2} = p/p^{\ominus} \tag{4.3}$$

Standard States for the Hydrogen Solute

There is no 'pure' counterpart corresponding to the atomic solute and an alternative standard state is required. According to applications, it is common practice to select one or the other of the following alternative standard states for a solute:[5-7]

1. A definite composition, usually a standard value of molality, $m_H^{\ominus}$, so that the activity, a_H of an arbitrary quantity of solute corresponding to a molality, m_H is:

$$a_H = m_H/m_H^{\ominus} \tag{4.4}$$

Unit molality is often selected as the standard value, $m_H^{\ominus}$, but it is not mandatory and some other value is sometimes more convenient.

2. A hypothetical standard state defined so that the activity of the solute equals its mole fraction at infinite dilution. i.e:

$$a_H \rightarrow x_H \text{ as } x_H \rightarrow 0 \tag{4.5}$$

In the composition range of interest, hydrogen solutions are so dilute that the solute exhibits Henrian activity and both scales are linear.

4.1.1.2 Pressure-Dependence of Solubility and Sievert's Isotherm

Selecting $p^{\ominus}$ and $m_H^{\ominus}$ as standard states, eqn 4.2 yields Sieverts' isotherm:[8]

$$m_H/m_H^{\ominus} = K^{\frac{1}{2}} \cdot (p/p^{\ominus})^{\frac{1}{2}} \tag{4.6}$$

where m_H is the molality of a solution in equilibrium with the gas phase at a pressure, p.

4.1.1.3 Temperature-Dependence of Solubility and the Van't Hoff Isobar

The change in Gibbs free energy, dG, for an infinitesimal change in the system is:

$$dG = \Sigma V dp - \Sigma S dT + \Sigma \mu dn \tag{4.7}$$

where V, S, μ and n are respectively the volumes, entropies, chemical potentials and numbers of moles of the participating species.

For isothermal isobaric transfer of dn_{H_2} moles of diatomic hydrogen from the gas phase to solution yielding $2dn_H$ moles of atomic solute ∂T and ∂p are zero and eqn 4.7 becomes:

$$
\begin{aligned}
dG &= -\mu_{H_2} dn_{H_2} + \mu_H dn_H \\
&= -\mu_{H_2} dn_{H_2} + 2\mu_H dn_{H_2} \\
&= -(\mu_{H_2}^{\ominus} + RT \ln a_{H_2}) dn_{H_2} + 2(\mu_H^{\ominus} + RT \ln a_H) dn_{H_2}
\end{aligned} \tag{4.8}
$$

where: $\mu_H^{\ominus}$, $\mu_{H_2}^{\ominus}$ are the chemical potentials of solute and gas in their standard states.

At equilibrium, $dG = 0$, so that:

$$-(\mu_{H_2}^{\ominus} + RT \ln a_{H_2}) + 2(\mu_H^{\ominus} + RT \ln a_H) = 0 \tag{4.9}$$

Rearranging:

$$\ln a_{\mathrm{H}} - \tfrac{1}{2}\ln a_{\mathrm{H}_2} = -\frac{\left(2\mu_{\mathrm{H}}{}^{\ominus} - \mu_{\mathrm{H}_2}{}^{\ominus}\right)}{2RT} \tag{4.10}$$

and since $2\mu_{\mathrm{H}}{}^{\ominus} - \mu_{\mathrm{H}_2}{}^{\ominus} = \Delta G^{\ominus}$ and $\Delta G^{\ominus} = \Delta H^{\ominus} - T\Delta S^{\ominus}$:

$$\ln a_{\mathrm{H}} - \tfrac{1}{2}\ln a_{\mathrm{H}_2} = -\frac{\Delta H^{\ominus}}{2RT} + \frac{\Delta S^{\ominus}}{2R} \tag{4.11}$$

The Van't Hoff Isobar in Terms of Solute Molality
Selecting $p^{\ominus}$ and $m_{\mathrm{H}}^{\ominus}$ as standard states for eqn 4.11 and converting to common logarithms yields the Van't Hoff isobar in the familiar form:

$$\log\left(\frac{m_{\mathrm{H}}}{m_{\mathrm{H}}{}^{\ominus}}\right) - \tfrac{1}{2}\log\left(\frac{p}{p^{\ominus}}\right) = -\frac{\Delta H^{\ominus}}{2.303 \times 2RT} + \frac{\Delta S^{\ominus}}{2.303 \times 2R} \tag{4.12}$$

where:

p is the prevailing pressure of the gas, referred to its pressure, $p_{\mathrm{H}}{}^{\ominus}$, in the selected standard state, usually 101 kPa (1 atm.).

m_{H} is the molality of the solute referred to a standard value, $m_{\mathrm{H}}{}^{\ominus}$, selected to lie within the range for which Henrian activity applies.

$\Delta H^{\ominus}$ and $\Delta S^{\ominus}$ refer to a solution in the standard state.

The Van't Hoff Isobar in Terms of Infinitely Dilute Mole Fraction
Selecting $p^{\ominus}$ and the infinitely dilute mole fraction, x_{H}, as standard states yields an alternative form of the equation needed for applications in Section 4.1.2.3 and 6.3.3.5:

$$\log x_{\mathrm{H}} - \tfrac{1}{2}\log\left(\frac{p}{p^{\ominus}}\right) = -\frac{\Delta H_{x_{\mathrm{H}}}^{\ominus}}{2.303 \times 2RT} + \frac{\Delta S_{x_{\mathrm{H}}}^{\ominus}}{2.303 \times 2R} \tag{4.13}$$

where the standard values, $\Delta H_{x_{\mathrm{H}}}^{\ominus}$ and $\Delta S_{x_{\mathrm{H}}}^{\ominus}$ are theoretical concepts with no physical reality because they refer to a *hypothetical* standard state for the solute.

4.1.2 SOME FUNDAMENTAL ASPECTS OF HYDROGEN SOLUTIONS

Electronic, statistical and crystallographic approaches deal with aspects of hydrogen solutions in metals that classical thermodynamics does not embrace. The features they reveal are essential to account for both the nature of the solutions and some effects on the metal that cannot otherwise be explained. Rigorous validation of the electronic and statistical principles is the province of specialists but the basic ideas and their implications are straightforward. The following sections introduce them, with references to comprehensive treatments.

4.1.2.1 Electronic Approach to the Enthalpy of Solution

The derivation of enthalpies of solution from first principles for multi-particle systems comprising hydrogen or other interstitial atoms dissolved in a host metal lattice can pose difficult mathematical problems, e.g. the application of band theories is frustrated by the expense of computer capacity and the requirement for periodicity.[9]

These problems can be partly circumvented using a semi-empirical approach, pioneered by Stott, Nørskov, Puska, Daw and their co-workers,[10–18] in which the solute atoms are considered to interact only with the delocalised electrons in the metal, ignoring its crystallographic structure. Calculation shows that a small solute atom accommodated interstitially in a metal host, with its acquired electron screening, is a compact object that can be treated as an quasi-atom whose energy is determined primarily by the prevailing electron density in the host. The real host is treated as if it were a spherically symmetrical uniform electron gas whose density equals that in the real host averaged over the region occupied by the quasi-atom; the concept is known as the *uniform density approximation* (UDA). The periodic positive potentials in the lattice are considered to serve only to restrain the electron density to its nominal value. To a first approximation, the parameter characterising the host is therefore its electron density, i.e. the electron population in a specified volume, $n_e.V^{-1}$ which can be computed by linear superposition of the electron densities of the constituent atoms.[19] Determining the energy of any particular solute species in any host is thus reduced to computing the energy, $\Delta E_{(hom)}$, required to insert it into a paramagnetic homogeneous electron gas as a function of the electron density. The approach can be extended to evaluate solute binding energy to features such as lattice vacancies, dislocations and surfaces that can be modelled as regions with diminished electron density.

The detailed calculations are specialised and are concerned principally with maintaining self-consistency and applying second order corrections. The relationships between electron density and the energies needed to implant elements of low atomic number have been computed and confirmed by independent investigators.[10, 16]

Calculations for hydrogen and helium are compared in Figure 4.1 to illustrate the origin of an attractive interaction for hydrogen.[10, 16] Helium has a filled 1s shell and the energy to insert the paired valency electrons at the Fermi level in the metal rises monotonously as the electron density rises from zero, so that the interaction is repulsive in *all* electron environments. In contrast, the energy to implant hydrogen at first falls to more negative values as the electron density rises from zero, reaching a minimum at a critical value of electron density and subsequently rising continuously, ultimately becoming positive. The reduction in energy in environments with low electron densities is because energy expended to promote the unpaired electrons from the 1s shell of the hydrogen atoms to the Fermi level in the metal is more than compensated by energy lost by relaxation of delocalised electrons from the metal back into the 1s shell, improving the electron screening, i.e. the neutral atoms assume some of the character of H⁻ ions, absorbing potential energy and the interaction is attractive. As the electron density rises further, the energy to promote electrons to the Fermi level becomes too high to be compensated and the interaction is repulsive.

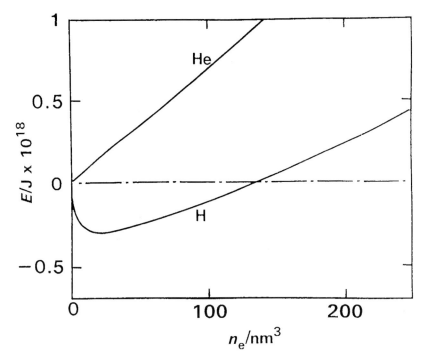

Fig. 4.1 Energy, E, required to insert atoms into an electron environment as a function of electron density, n_e, showing attractive interaction for hydrogen in low electron densities and repulsive interaction for helium in all electron environments. (Stott and Zaremba,[10] Puska, Niemenin and Maninen).[16] Values from original references are converted to SI units.

Average electron densities in metals[10, 15, 16] are typically in the range 100 to 200 nm^{-3}, which is only partly within the range, 0 to 135 nm^{-3}, for which the enthalpy to insert hydrogen atoms is negative. However, electron densities in real metals are not uniform as assumed in the simple theory but range between maxima and minima[15] in phase with lattices sites. The enthalpies of solution for monatomic hydrogen in several metals, given in Table 4.1, derived by subtracting the enthalpy of disassociation from measured values for diatomic hydrogen are all negative, consistent with minima low enough to interact attractively with hydrogen atoms.

4.1.2.2 Statistical Approach to Solubility

As an alternative to the classical approach, equilibria for gas reactions can be formulated from partition functions which represent the statistical distribution of energy among the constituent atoms or molecules. The entropy change for dissolution of hydrogen in a metal is attributed to the difference in the partition functions for hydrogen in the gas

Table 4.1 Enthalpies of Solution for Hydrogen in Some Metals[*]

Metal	Standard Enthalpies, $\Delta H^{\ominus}$	
	Referred to Diatomic Hydrogen, kJ mol^{-1}	Referred to Monatomic Hydrogen, kJ mol^{-1}
Th	− 94.1	− 260.6
Zr	− 73.2	− 246.0
Ti	− 41.8	− 234.3
V	− 32.2	− 229.7
Pd	− 8.5	− 217.6
Mo	+ 14.6	− 206.3
Ni	+ 23.4	− 201.7
Fe	+ 29.3	− 198.7
Co	+ 30.5	− 198.3
Cu	+ 59.0	− 184.1
Pt	+ 148.1	− 139.3

[*]R.M. Barrer[20]

phase and as atoms dissolved in the metal. This treatment is necessary but insufficient because it treats the metal as a continuum, ignoring the crystal structure which introduces an additional contribution to the entropy change, i.e. entropy of mixing; this omission and its implications are considered in Section 4.1.2.3. The enthalpy change is determined by the difference in potential energy between a hydrogen molecule and its constituent atoms within the metal.

Dissolution of the diatomic gas can be represented as dissociation followed by solution, yielding an atomic solute:

$$H_2(g) = 2H(g)$$

$$\downarrow$$

$$2H(g) = 2H(\text{solution in metal}) \tag{4.14}$$

Equilibrium for dissociation of a diatomic gas is given in standard texts, e.g. Glasstone:[22]

$$\frac{n_H^2}{n_{H_2}} = \frac{(Q_H)^2}{Q_{H_2}} \times e^{\frac{-\chi_d}{kT}} \tag{4.15}$$

where: n_H, n_{H_2} are the numbers of atoms and molecules of hydrogen.

Q_H, Q_{H_2} are the partition functions for the monatomic and diatomic gases.

χ_d is the potential energy difference between a hydrogen molecule and its constituent atoms in the lowest quantum states.

and the standard expressions for the partition functions[22, 23] are:

Monatomic gas: $Q_H = \dfrac{(2\pi mkT)^{\frac{3}{2}} \times V \times g_1}{h^3}$ (4.16)

Diatomic gas: $Q_{H_2} = \dfrac{\{2\pi(2m)kT\}^{\frac{3}{2}} \times V \times g_2}{h^3} \times \dfrac{8\pi^2 IkT}{2h^2}$ (4.17)

where: m is the mass of a hydrogen atom.
        V is the volume of the phase.
        I is the moment of inertia of the hydrogen molecule.
        h, k are respectively Planck's and Boltzman's constants.
        g_1, g_2 are the statistical weight factors with values of 2 and 1 respectively for resultant electron spin, i.e. for single electrons in the atoms and paired electrons in the molecules.

To extend the treatment to include solution of the atoms in the metal, a partition function is needed for the solute. This is not straightforward because the motions of the solute are uncertain. Hydrogen is very mobile in metals and the partition function for the solute may have contributions from translation in the metal and oscillation within interstitial sites. Fowler and Smithells[24] made the simplifying assumption that the solute exists as quasi-free atoms each with potential energy $(\chi_s + W)$ less than that of an atom in the gas phase in the lowest quantum state, where χ_s is the minimum value and $W \geq 0$. With this assumption the partition function for the solute, Q_s, is:

$$Q_s = \frac{(2\pi mkT)^{\frac{3}{2}} \times V \times g_3}{h^3}$$ (4.18)

As described in Section 4.1.2.1, the solute species is viewed as quasi-atoms interacting by accepting electrons from the metal into their 1s orbitals; the resultant electron spin is not fully cancelled because on average the orbital is not completely filled so that the g_3 is > 1 but ≪ 2. With this reservation, as a reasonable simplifying assumption it is set equal to 1.

Equilibrium between atoms in the gas phase and atoms in solution is given by:

$$\frac{n_s}{n_H} = \frac{Q_s}{Q_H} \times e^{\frac{-\chi_s}{kT}} \times e^{\frac{-W}{kT}}$$ (4.19)

where: n_H is the number of atoms in the gas.
        n_s is the number of solute atoms in volume, V, of the metal.

The exponential containing W in eqn 4.19 expresses the statistical distribution of the energies of solute atoms within the volume of the metal. For a dilute solution the equation is simplified because values of W are so small that the exponential approaches unity.

Combining eqns 4.15 and 4.19, eliminates n_H and substituting for Q_s and Q_{H_2} yields:

$$n_s = \frac{n_{H_2}^{1/2} \times Q_s}{Q_{H_2}^{1/2}} \times e^{-\frac{(\chi_s + 1/2\chi_d)}{kT}} = \frac{n_{H_2}^{1/2} \times V^{1/2} \times m^{3/4}}{2\pi^{1/4} \times h^{1/2} \times I^{1/2}} \times e^{-\frac{(\chi_s + 1/2\chi_d)}{kT}}$$ (4.20)

Substituting for n_{H_2} using the ideal gas law ($pV = n_{H_2}kT$) and rearranging, yields for the concentration, ns/V, of solute atoms in the metal:

$$\frac{n_s}{V} = p^{1/2} \times \frac{m^{3/4}}{2(\pi kT)^{1/4} \times (hl)^{1/2}} \times e^{-\frac{(\chi_s - 1/2\chi_d)}{kT}}$$ (4.21)

The pre-exponential factor is insensitive to the $T^{1/4}$ term and for a restricted temperature range the equation is simplified by considering it constant. Converting solute concentration, n_s/V, to molality, $m_H/m_H^{\ominus}$, referring pressure to a standard value, $p^{\ominus}$, taking logarithms and collecting constants into a single term, C, yields an equation similar to eqn 4.12:

$$\ln\frac{m_H}{m_H^{\ominus}} - \frac{1}{2}\ln\frac{p}{p^{\ominus}} = \frac{-\Delta H^{\ominus}}{2RT} + C$$ (4.22)

since: $\Delta H^{\ominus} = L(\chi_s - \frac{1}{2}\chi_d)$ (4.23)

and: $R. = kL$, where L is Avogadro's constant. (4.24)

4.1.2.3 Crystallographic Aspects of Hydrogen Solutions

Thermodynamic, electronic and quantum statistical approaches to the solution of hydrogen in metals are incomplete to the extent that they ignore the crystal structure of the metal. The structure determines the accommodation available to the solute and contributes a configurational term, the entropy of mixing, to the entropy change on dissolution of the gas. The following discussion considers this aspect of hydrogen-metal systems and shows that the entropy of solution for hydrogen is the same for all metals with the same uniform crystallographic form when compared on a common basis. This principle is used to formulate a general equation for solutions of hydrogen in FCC metals that is useful in resolving some practical problems that emerge in Chapter 6, i.e.:

1. Application of the Van't Hoff isobar to measured values of solubility.
2. Validation of anomalous solubilities of hydrogen in aluminium–lithium alloys.

Accommodation for Hydrogen Solute Atoms in FCC Metals

Consider the accommodation available in FCC metals. The inscribed radii of tetrahedral and octahedral interstices calculated from the spacing of the atoms in FCC metals are typically in the ranges 0.03– 0.04 nm and 0.05– 0.07 nm respectively, e.g. 0.032 nm and 0.059 nm for aluminium.[2] Comparing these radii with the size of the hydrogen atom, usually represented by the Bohr radius, 0.053 nm identifies the octahedral interstices as the only option for accommodating the solute without the penalty of severe distortion.

Entropy of Mixing

The entropy of mixing of a component in a solid system, ΔS_m, is determined by the statistical distribution of its atoms among available sites, given by standard equations:[25-27]

$$\Delta S_m = -R\left(\frac{n}{N}\right)\ln\left(\frac{n}{N}\right) \tag{4.25}$$

where n is the number of atoms and N is the number of sites.

The number of octahedral sites in FCC metals, N, equals the number of metal atoms, N_M so that for dilute solutions, the ratio of the number of hydrogen solute atoms, n to the number of available sites is virtually equal to the mole fraction of solute, x_H, i.e.:

$$\frac{n}{N} \rightarrow \frac{n}{N_M} = x_H \text{ as } n_H \rightarrow 0 \tag{4.26}$$

Substituting x_H for n/N in eqn 4.25:

$$\Delta S_m = -Rx_H \ln x_H \tag{4.27}$$

Thus when the quantity of the solute is expressed as a mole fraction, x_H, the entropy of mixing for hydrogen in a regular FCC lattice has a characteristic value, independent of the particular metal considered.

Standard Entropy of Solution in FCC Lattices

The standard entropy change for dissolution of diatomic hydrogen gas in a metal referred to one mole of monatomic solute, $\Delta S^{\ominus}$, is the sum of three components:

$$\Delta S^{\ominus} = \frac{1}{2}\Delta S_d^{\ominus} + \Delta S_p^{\ominus} + \Delta S_m^{\ominus} \tag{4.28}$$

where: $\Delta S_d^{\ominus}$ is the standard entropy of dissociation of diatomic to monatomic gas.

$\Delta S_p^{\ominus}$ is the standard entropy change for the difference in partition functions for atoms in the gas phase and solute atoms in the metal.

$\Delta S_m^{\ominus}$ is the standard entropy of mixing for the solute atoms in the metal.

Since $\Delta S_d^{\ominus}$ and $\Delta S_p^{\ominus}$ are characteristics of the gas and $\Delta S_m^{\ominus}$ is characteristic of the lattice, the total entropy of solution, $\Delta S^{\ominus}$, given by eqn 4.28 is determined only by the crystallographic structure of the host metal. Thus when the activity of the solute is referred to the infinitely dilute mole fraction, x_H, the standard entropy of solution for hydrogen in a regular FCC lattice has a characteristic value, applicable to all FCC metals.

With this information, the entropy term in eqn 4.13, given in Section 4.1.1.3 can be standardised, yielding a general equation for solutions of hydrogen in FCC metals. This is done graphically, using the measured values of the solubility for nickel,[28, 29] γ-iron,[29] cobalt,[28, 29] copper,[31, 32] platinum[33] and aluminium[31, 34] given in Table 4.2 as mole fractions.

Table 4.2 Solubilities of Hydrogen in FCC Metals at 101 kPa (1 atm.) - Mole Fractions, x_H.

Temperature		Nickel[28, 29]	γIron[29]	Cobalt[28, 29]	Copper[31, 32]	Platinum[35]	Aluminium[31, 34]
°C	$10^4 K/T$	$x_H \times 10^4$	$x_H \times 10^4$	$x_H \times 10^4$	$x_H \times 10^5$	$x_H \times 10^5$	$x_H \times 10^7$
300	17.45	1.05	–	–	0.06	–	–
400	14.86	1.31	–	–	0.28	–	–
470	13.46	–	–	–	–	–	0.96
500	12.94	1.73	–	–	0.62	–	1.45, 1.93
530	12.45	–	–	–	–	–	2.65, 2.89
560	12.01	–	–	–	–	–	4.34
590	11.58	–	–	–	–	–	5.78
600	11.46	2.25	–	0.47	1.19	–	–
700	10.28	2.93	–	0.64	1.53	–	–
800	9.32	3.67	–	0.97	3.01	1.7	–
900	8.53	4.45	2.37	1.32	5.05	–	–
1000	7.86	5.24	2.73	1.74	7.60	3.3	–
1100	7.28	6.03	3.23	2.27	–	5.9	–
1200	6.79	6.81	3.74	2.85	–	9.4	–
1250	6.57	–	4.24	–	–	–	–
1300	6.36	7.34	4.70	3.53	–	13.0	–
1400	5.98	8.65	–	4.08	–	–	–

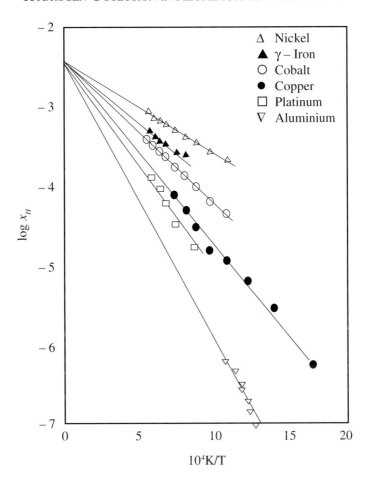

Fig. 4.2 Van't Hoff isobars at 101325 Pa (1 atm.) for hydrogen solutions in some FCC metals with the solute activity referred to the infinitely dilute mole fraction standard state, $a_H \to x_H$ as $x_H \to 0$. All of the isobars converge to a common intercept at $K/T = 0$ with the value $[\log x_H]_{K/T=0} = -2.40$.

For a prescribed standard pressure, e.g. $p = p^{\ominus} = 101$ kPa, the equation reduces to:

$$\log x_H = -\frac{\Delta H^{\ominus}_{x_H}}{2.303 \times 2RT} + \frac{\Delta S^{\ominus}_{x_H}}{2.303 \times 2R} \tag{4.29}$$

Figure 4.2 gives isobars for these metals in the format of eqn 4.29, extrapolated to intercept the abscissa at $K/T = 0$, where the enthalpy term vanishes and $\log x_H$ is a function only of the entropy of solution. The extrapolations for all of the FCC metals converge to the same intercept. This corresponds to the entropy term in eqn 4.29 and its numerical value read from Figure 4.2 is:

$$\left[\log x_H\right]_{\frac{K}{T}=0} = -2.40 \tag{4.30}$$

Substituting in eqn 4.29 yields the required general equation for the solubility of hydrogen occupying the octahedral interstices in *any* pure FCC metal:

$$\log x_H - \tfrac{1}{2}\log\left(\frac{p}{p^{\ominus}}\right) = -\frac{\Delta H^{\ominus}_{x_H}}{2.303 \times 2RT} - 2.40 \tag{4.31}$$

It is interesting to observe that the differences in solubility of hydrogen in various FCC metals depend only on the standard enthalpy $\Delta H^{\ominus}_{x_H}$ as assumed in the quantum statistical approach given in Section 4.1.2.2.

Analogous equations can be written for metals with other crystal forms, e.g. BCC, HCP or liquids if the sites suitable for occupation by the solute can be identified.

4.2 SPONTANEOUS BREAKDOWN OF SUPERSATURATED SOLUTIONS

For the reasons given in Chapter 8, the quantities of hydrogen in most manufactured aluminium and aluminium alloy products are much higher than the very low solubilities in the solid metal in equilibrium with hydrogen gas at atmospheric pressure and if it were all dissolved, it would produce a solution with a very high hydrogen activity. However, at elevated temperatures, hydrogen is mobile in the metal and the energy of the system can be reduced by separation of an internal gas phase in equilibrium with a residual solution. The ensuing breakdown of the solution can generate microporosity which damages the metal if its development is uncontrolled, as described in Chapter 9. The gas phase separates as spherical micropores nucleated heterogeneously rather than homogeneously because aluminium and aluminium alloy products usually contain spherical nuclei, 1–2 µm in diameter disseminated throughout their structures,[35-38] as also explained in Chapter 9.

The chemical potential difference driving the nucleation and growth of micropores filled with hydrogen is opposed by their surface energy. There is an important difference between nucleation of a gas phase and the nucleation of solid precipitates described by standard nucleation theory; a gas phase is compressible and the pressure in the system is not constant but falls progressively as the pores expand so that equilibria correspond with minima in the Helmholtz function, A.

The breakdown of solutions of hydrogen in aluminium can be characterised as follows in Section 4.2.1, using the approach developed by Bondarenko and Khmelinen[39] for the separation of gaseous fission products from metals and dissolved oxygen from silicon.

4.2.1 NUCLEATION AND GROWTH OF HYDROGEN–FILLED MICROPORES

Consider a closed system comprising a mass of the metal containing n moles of hydrogen distributed as n_H moles of atomic solute and n_{H_2} moles of diatomic gas in N spherical pores:

$$n = n_H + 2n_{H_2} \qquad (4.32)$$

The Helmholtz function, A, is the sum of contributions, A_H from the solute, A_{H_2} from the gas phase, A_σ from surface energy and A_ε from strain energy. The increase, dA, for isothermal transfer of an infinitesimal quantity of solute, dn_H moles, to any one pore, i, yielding dn_{iH_2} moles in the gas phase, is:

$$dA = dA_H + dA_{iH_2} + dA_{i\sigma} + dA_{i\varepsilon} \qquad (4.33)$$

In standard theory, the term, $dA_{i\varepsilon}$, represents elastic strain energy imposed by volume changes accompanying phase changes at low temperatures but at the temperatures at which hydrogen is mobile in aluminium, the metal is plastic so that strain energy is relaxed and $dA_{i\varepsilon} = 0$. The solute volume, V_H, is constant because it equals the metal volume so that $dV_H = 0$ and $dn_H = -2dn_{iH_2}$ from eqn 4.1. Therefore the terms in eqn 4.33 are:

$$dA_{i\varepsilon} = 0, \qquad (4.34)$$

$$dA_H = -p \cdot dV_H + \mu_H \cdot dn_H = 0 - 2\mu_H \cdot dn_{H_2} \qquad (4.35)$$

$$dA_{iH_2} = -p \cdot dV_{iH_2} + \mu_{iH_2} \cdot dn_{iH_2} = -p \cdot 4\pi r^2 \cdot dr + \mu_{iH_2} \cdot dn_{iH_2} \qquad (4.36)$$

$$dA_{i\sigma} = \sigma \cdot d(4\pi r^2)dr = 8\sigma\pi r \cdot dr \qquad (4.37)$$

where: μ_H, μ_{iH_2} are the chemical potentials of hydrogen atoms and molecules.
V_{iH_2}, r are the pore volume and radius.
p is the pressure in the internal gas phase.
n_{iH_2} is the number of hydrogen molecules in the pore.
σ is the surface tension coefficient for the metal.

Substituting these terms in eqn 4.33, assuming Henry's law for the solute and ideality for the gas:

$$dA = -2\mu_H \cdot dn_{iH_2} - p \cdot 4\pi r^2 \cdot dr + \mu_{iH_2} \cdot dn_{iH_2} + 8\sigma\pi r \cdot dr$$

$$= -2\left(\mu_H^{\ominus} + RT\ln\frac{n_H}{n_H^{\ominus}}\right)dn_{iH_2} - p \cdot 4\pi r^2 \cdot dr +$$

$$\left(\mu_{iH_2}^{\ominus} + RT\ln\frac{p}{p^{\ominus}}\right)dn_{iH_2} + 8\sigma\pi r \cdot dr \qquad (4.38)$$

where:
$p/p^{\ominus}$ is pressure of the gas in the pore referred to its pressure in a standard state, $p^{\ominus}$, e.g. 101.325 kPa,
n_H, $n_H^{\ominus}$ are the numbers of moles of solute in equilibrium with p and $p^{\ominus}$.

$\mu_H^{\ominus}$, $\mu_{iH_2}^{\ominus}$ are respectively the chemical potentials of the solute hydrogen and the gas phase in their standard states, n° and $p^{\ominus}$.

Since n_{iH_2} and r are independent variables, there are two conditions for equilibrium:

$$\left[\frac{dA}{dr}\right]_{n_{iH_2}} = 0 \tag{4.39}$$

$$\left[\frac{dA}{dn_{iH_2}}\right]_r = 0 \tag{4.40}$$

From eqn 4.39: $-p \times 4\pi r^2 + 8\sigma\pi r = 0$ \hfill (4.41)

and rearranging: $p = \dfrac{2\sigma}{r}$ \hfill (4.42)

From eqn 4.40: $-2\left[\mu_H^{\ominus} + RT\ln\left(\dfrac{n_H}{n_H^{\ominus}}\right)\right] + \left[\mu_{iH_2}^{\ominus} + RT\ln\left(\dfrac{p}{p^{\ominus}}\right)\right] = 0$ \hfill (4.43)

and rearranging: $n_H = n_H^{\ominus} \times [\exp -(\mu_H^{\ominus} - \frac{1}{2}\mu_{iH_2}^{\ominus})/RT] \times (p/p^{\ominus})^{\frac{1}{2}}$ \hfill (4.44)

which is an alternative form of Sievert's isotherm, given in eqn 4.6.

Pores are identical at equilibrium, so that eqns 4.43 and 4.44 extend to the whole system. The quantity of hydrogen in the internal gas phase n_{H_2}, is given by applying the ideal gas law to the total pore volume, $V_{iH_2} \times N$, substituting for p from eqn 4.42:

$$n_{H_2} = \frac{pV_{iH_2} \times N}{RT} = \frac{2\sigma}{r} \times \frac{\left[\frac{4}{3} \times \pi r^3 \times N\right]}{RT} = \frac{8\pi N\sigma}{3RT} \times r^2 \tag{4.45}$$

Substituting for n_H and n_{H_2} in eqn 4.32 from eqns 4.44 and 4.45 yields a relation between the hydrogen content of the system, n and the number, N, and radius, r, of the pores:

$$n = [(n_H^{\ominus} \times 2^{\frac{1}{2}} \times \sigma^{\frac{1}{2}})(p^{\ominus})^{-\frac{1}{2}} \times \exp -(\mu_H^{\ominus} - \frac{1}{2}\mu_{iH_2}^{\ominus})/RT] \times r^{-\frac{1}{2}} + [16\pi N\sigma/3RT] \times r^2$$

$$= X \times r^{-\frac{1}{2}} + Y \times Nr^2 \text{ (collecting constants into coefficients X and Y)} \tag{4.46}$$

Differentiation discloses a minimum, at the co-ordinates, (r', n'), where:

$$r' = \left[\frac{X}{4NY}\right]^{\frac{2}{5}}, \quad n' = X \times \left[\frac{X}{4Y}\right]^{-\frac{1}{5}} + Y \times \left[\frac{X}{4NY}\right]^{\frac{4}{5}} \tag{4.47}$$

Figure 4.3 (a) is a schematic plot of eqn 4.46, illustrating the following features:
1. If $n < n'$, the equation has no roots, i.e. the hydrogen solution is unsaturated so that any internal gas phase is unstable.
2. If $n > n'$, there are two roots, r_1 and r_2, corresponding with extrema in the Helmholtz function, A, illustrated schematically in Figure 4.3b. The smaller root, r_1,

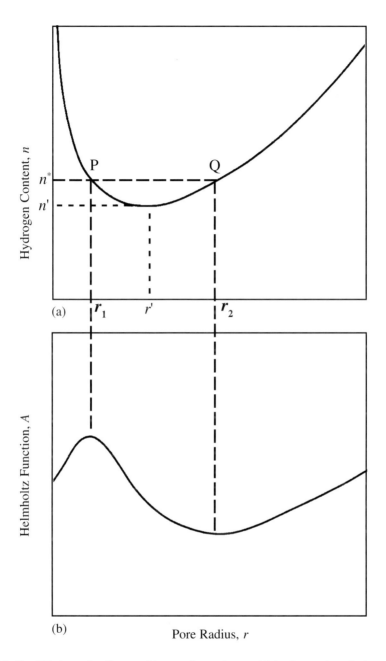

Fig. 4.3 Equilibrium of a dispersed internal gas phase with hydrogen in solution. Schematic plots based on eqn 4.46.
(a) Hydrogen content, n, versus equilibrium pore radius, r, exhibiting a minimum at (r', n') and (b) Helmholtz function, A, versus pore radius, r, for a hydrogen content, n^* above n'. The maximum and minimum correspond with the conjugate roots, P and Q, of eqn 4.46.

corresponding with a maximum in A, is the classical critical radius for nucleation; the larger root, r_2, corresponding with a minimum, identifies a stable state.

4.2.3 PORE STABILITY DIAGRAM

Constructing a tangent to the minimum in Figure 4.3a, yields the diagram, given in Figure 4.4, with domains that predict the behaviour of pores, if kinetics allow, as follows:

1. Pores with system co-ordinates (n, r) below AMC are unstable and are annihilated.
2. Pores with co-ordinates lying in BMC contract to a stable size at a point on MB.
3. Pores with co-ordinates lying in AMB expand to a stable size at a point on MB.
   The diagram has applications relating to various effects observed in production processes.

4.3 HYDROGEN TRAP SITES

Aluminium and aluminium alloys contain structural features on macroscopic, microscopic and atomic scales that offer preferred sites for hydrogen, collectively known as hydrogen traps. It is convenient to assign them to the following categories:

1. Diatomic traps, where an internal gas phase collects in small pores or pockets.
2. Monatomic traps, where solute atoms are selectively bound to point, line or planar lattice defects.
3. Chemical traps, comprising components or impurities with affinities for hydrogen that can immobilise it by chemical reaction.

4.3.1 DIATOMIC TRAPS

4.3.1.1 Microporosity
Microporosity formed by the decomposition of hydrogen solutions described in Section 4.2 are diatomic traps in equilibrium with the residual solution. The pore nuclei that are the precursors of the traps are common features in manufactured semi-finished forms of aluminium and its alloys.[35–38] If undisturbed, they have little effect but if allowed to develop as described in Chapters 8, 9 and 10 they can compromise the integrity of the metal.

4.3.1.2 Extraneous Artifacts
Flaws introduced by careless or uninformed manufacturing practices offer large traps for diatomic hydrogen. Examples include:

1. Remnants of gross porosity or oxide fragments in metal fabricated from ingots cast from inadequately prepared liquid metal.[35]

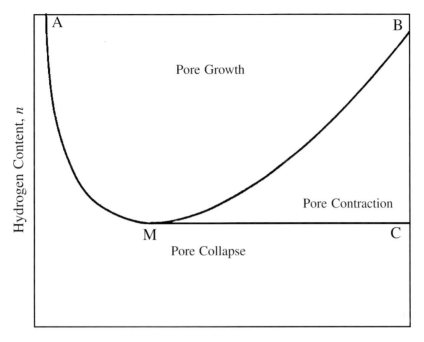

Fig. 4.4 Diagram based on Figure 4.3a, showing domains in which the co-ordinates of hydrogen content, n, and pore radius, r predict pore growth, contraction or collapse.

2. Defects generated mechanically during fabrication, including cracks and incomplete welding at roll-bonded core/cladding interfaces.

Such artifacts were once responsible for outbreaks of serious hydrogen-related damage to critical products but their origins and methods of eliminating them are now so well understood that they are rarely encountered in products supplied by reputable manufacturers. Further discussion is deferred to Chapter 10, where these artifacts are reviewed in their appropriate practical context.

4.3.2 MONATOMIC TRAPS

The preferential attachment of hydrogen atoms in metallic solutions to lattice defects is well-known and it is covered in general reviews of the association of solute atoms with structural features.[40, 41]

Binding energies between solute atoms and single vacancies are accessible by e.g. band and pair-potential theories[9, 42–47] but because they cannot cope with regions where

Table 4.3 Binding Energies for Helium-Vacancy and Hydrogen-Vacancy Pairs[*]

Valency of Vacancy	Electron Density nm^{-3}	Values for Binding Energies, 10^{19} E/J			
		Helium-Vacancy Pair		Hydrogen-Vacancy Pair	
		Electron Density Calculation	Classical Calculation	Electron Density Calculation	Classical Calculation
1	200	− 7.9	− 9.7	− 3.7	−5.3
1	59	− 3.1	− 3.6	-	-
1	47 (≡ Li)	-	-	− 0.4	0.0
1	26 (≡ Na)	-	-	+ 0.2	+1.0
2	200	− 10.1	− 11.3	-	-
2	86 (≡ Mg)	-	-	− 1.4	−1.6
2	59	− 3.7	− 4.8	-	-
3	180 (≡ Al)	-	-	−4.2	−5.0

[*]M.J. Stott and E. Zaremba.[10]

the host volume is ambiguous, these approaches lack the flexibility to deal with the segregation of hydrogen to extended defects,[17] e.g. dislocations, crack roots and surfaces, which is the source of many technically important effects.

The problem can be resolved by extending the approach based on the interaction of solute hydrogen atoms with the delocalised electrons in the metal, as described in Section 4.1.2.1 because it does not need information on the local host volume. The presence of a lattice defect attenuates the electron density in its immediate vicinity, changing the binding energy to interstitial solute atoms and they respond accordingly. The normal interstitial electron densities for most metals[10, 16] lie within the range 100–200 nm^{-3}, which is above the optimum value, 17 nm^{-3}, for which an implanted hydrogen atom has the minimum energy.[10, 16] Consequently, as the normal density diminishes towards vacancies, dislocations, interfaces or free surfaces, it approaches or passes through the optimum value, yielding negative binding energies that offer hydrogen trap sites.

Stott and Zaremba[10] represent the influence of a vacancy on the local electron environment by envisaging it as a "hole" in the positive background charge with the same valency and volume as the missing metal atom. Table 4.3 shows reasonably good agreement between their calculated values of binding energies for helium and hydrogen atoms to the attenuated electron density and corresponding values for binding to lattice vacancies determined by e.g. the Kohn-Sham[48] method. The positions of hydrogen atoms bound to vacancies can also be located, because the maximum binding energy corresponds with the position where the electron density is nearest to the optimum. For example, Nørskov et al.[15] and Myers et al.[49] have both shown that a hydrogen atom bound to a vacancy in iron or nickel is displaced towards an adjacent octahedral site.

In principle, the free surface of a metal is a multiple trap site, where hydrogen atoms seek out the optimum electron density as it is attenuated to a vanishingly small value. It is an application where electron density calculations circumvent the need to define a surface plane. As examples, Daw and Baskes[17] assessed the theoretical distribution of hydrogen on clean nickel and palladium surfaces as functions of the outcropping orientation. Application of the concept to real metal surfaces is less obvious, because they are almost invariably covered by air-formed oxide films and trapping is determined by electron densities prevailing at the metal/oxide interface and oxide/atmosphere interfaces.

Hydrogen is expected to bind less strongly to dislocations, self-interstitial atoms and grain boundary disorder than to vacancies because the dilution of the electron density is less.

4.3.3 CHEMICAL TRAPS

Chemical trapping is not usually a consideration for aluminium and its alloys since neither the pure metal nor the common alloying elements form stable hydrides under conditions of metallurgical significance. There are only two elements relevant to manufactured alloys that need be considered, the strongly alkaline metals, lithium[50-52] and sodium[53] and even for these elements, interest is limited to the specialised applications described in Chapter 10.

Lithium is a component of some specialised multicomponent age-hardening aerospace alloys, present typically in the range 2.0 to 2.5 mass%. The partial phase diagram for the aluminium-lithium binary system due to McAlister,[54] reproduced as Figure 6.18 in Chapter 6, shows that at above about 450°C (723 K) these compositions are within the α–phase.

Sodium is a contaminant of aluminium. It is absorbed from the electrolyte, cryolite, $NaAlF_6$[58] in the Hall-Herault process and without treatment to remove it, 0.002 to 0.003 mass% sodium typically persists through alloying and casting to be retained in the solidified metal.

4.3.3.1 The Stability of Lithium Hydride in Aluminium-Lithium Alloys

The standard Gibbs free energy for LiH formation in the temperature range 452–972 K is:[55]

$$Li + \tfrac{1}{2}H_2 = LiH \tag{4.48}$$

$$\Delta G^{\ominus} = -95186 + 83.5\,T \;\; J\,mol^{-1} \tag{4.49}$$

Lithium activity is not known throughout the whole of the α–phase field but at the $\alpha/(\alpha + \beta)$ phase boundary it is given by Veleckis' equation:[56]

Table 4.4 Hydrogen Activities, a_{H_2}, in Equilibrium with Lithium Hydride at the $\alpha/(\alpha + \beta)$ Phase Boundary in Binary Aluminium – Lithium Alloys

Temperature		Boundary Compostion Lithium, Mass Fraction[*]	Hydrogen Activity a_{H_2}
°C	T/K		
602[†]	875	3.7×10^{-2}	1.97
560	833	3.5×10^{-2}	1.00
500	773	2.8×10^{-2}	0.32
450	723	2.5×10^{-2}	0.11
400	673	2.1×10^{-2}	0.03

[*]McAlister[54]   [†]Eutectic Temperature

$$\ln(a_{Li}) = 2.662 - \frac{5302}{T} \qquad (4.50)$$

Applying the Van't Hoff isobar for equilibrium along the boundary:

$$\Delta G^{\ominus} = -RT \ln \frac{a_{LiH}}{a_{Li} \times a_{H_2}^{1/2}} \qquad (4.51)$$

Substituting for $\Delta G^{\ominus}$ from eqn 4.49 and for $\ln(a_{Li})$ from eqn 4.50, and assuming LiH is pure, so that $a_{LiH} = 1$:

$$\ln a_{H_2} = -\frac{12285}{T} + 14.75 \qquad (4.52)$$

Eqn 4.52 yields the values of hydrogen activity in equilibrium with lithium hydride at selected intervals along the $\alpha/(\alpha + \beta)$ phase boundary given in Table 4.4. These values indicate that for unit hydrogen activity (101 kPa) and temperatures < 833 K (560°C) the lithium activities along most of the $\alpha/(\alpha + \beta)$ phase boundary are sufficient to stabilise the hydride. This is consistent with Aronson and Solzano's evidence[57] that two-phase mixtures of conjugate α and β phases are in equilibrium with the hydride at hydrogen pressures < 101 kPa in the temperature range 723–823 K. More information would be needed to identify a domain in the α phase within which the hydride is expected to be stable. Furthermore, a hydride phase does not necessarily nucleate when the activities of lithium and hydrogen are sufficient to stabilise it, because factors associated with clustering in the α–phase can intervene as explained in Chapter 6, Section 6.3.3.4.

4.3.3.2 The Stability of Sodium Hydride in Aluminium and its Alloys

Sodium is almost insoluble in solid aluminium, e.g. the solubility is < 0.0005 mass% at 400°C (673 K) and almost all of any sodium in the metal is present in a grain boundary

Table 4.5 Hydrogen Activities, a_{H_2}, in Equilibrium with Sodium Hydride and Pure Elemental Sodium as a Function of Temperature

Temperature	°C	550	500	450	422	400	350
	T/K	823	773	723	695	673	623
Hydrogen Activity,	a_{H_2}	22.9	7.6	2.2	1.0	0.5	0.1

phase.[53] Conditions for formation of sodium hydride depend on the composition of this phase, which is determined by the natures and quantities of other components of the metal.

The standard Gibbs free energy for NaH formation in the temperature range 452–972 K is:[55]

$$Na + \tfrac{1}{2}H_2 = NaH \tag{4.53}$$

$$\Delta G^{\ominus} = -58390 + 84T \text{ J} \tag{4.54}$$

Applying the Van't Hoff isobar:

$$\Delta G^{\ominus} = -RT \ln \frac{a_{NaH}}{a_{Na} \times a_{H_2}^{1/2}} \tag{4.55}$$

Substituting for $\Delta G^{\ominus}$ from eqn 4.54, assuming $a_{Na} = a_{NaH} = 1$ and rearranging gives an expression for the equilibrium hydrogen activity as a function of temperature:

$$\ln a_{H_2} = \frac{-14046}{T} + 20.2 \tag{4.56}$$

Eqn 4.56 yields the values given in Table 4.5.

In high purity aluminium, the grain boundary phase is pure sodium that can be converted to sodium hydride *in situ* if the metal is exposed to hydrogen activities higher than the values given in Table 4.5, yielding microscopic cubic bodies as illustrated in Figure 4.5.

In commercial grades of pure aluminium e.g. Alloy AA 1100 and alloys based on it with < 2 mass% magnesium, the activity of silicon is high enough to convert the grain boundary phase to a compound, $NaAlSi_x$, which resists conversion to hydride. However in alloys with higher magnesium contents, the precipitation of Mg_2Si[60] reduces the silicon activity below that needed to stabilise $NaAlSi_x$, so that the grain boundary phase reverts to pure sodium[53] that can be converted to sodium hydride when exposed to hydrogen, in the same way as for high purity aluminium.

These phenomena have been verified experimentally by Ransley and Talbot.[53] Samples of alloys with 0.15% silicon and various magnesium contents and containing sodium were heated in hydrogen at 1 atm pressure and 400°C until the reaction was complete.

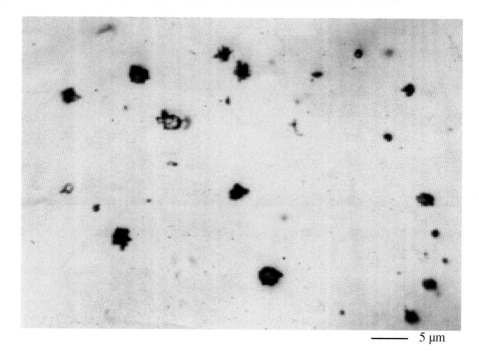

——— 5 μm

Fig. 4.5 Micrograph of an experimental aluminium – 5% magnesium alloy containing 0.05% sodium after heating in hydrogen at 450°C, showing discrete sodium hydride particles. (Ransley and Talbot).[53]

According to Table 4.5, these conditions are suitable to stabilise sodium hydride. The absorbed hydrogen was measured by vacuum extraction,[61] described later in Chapter 5, reporting the results as hydrogen contents as defined below in Section 4.4.2. These results are plotted in Figure 4.6, which shows that alloys with < 2 mass % magnesium absorbed only small quantities of hydrogen but alloys with higher magnesium contents absorbed sufficient hydrogen to account for conversion of most of the sodium to sodium hydride, as indicated by the line tracing out the stoichiometric ratio of sodium to hydrogen.

4.4 QUANTITATIVE DESCRIPTION OF OCCLUDED HYDROGEN IN INDUSTRY

4.4.1 DETERMINATE AND INDETERMINATE OCCLUDED SYSTEMS

The theoretical framework needed to interpret the behaviour of hydrogen in aluminium was developed on the basis of idealised systems in which the quantities of hydrogen in

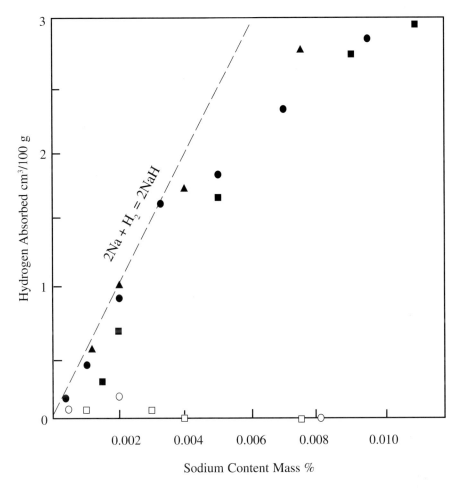

Fig. 4.6 Effect of magnesium content on sodium hydride formation in alloys with 0.15 mass% silicon assessed from the hydrogen content absorbed from the gas at 450°C and 101 kPa. Dashed line represents NaH stoichiometric ratio.

●   5   mass% magnesium + 0.15 mass% silicon
▲   3   mass% magnesium + 0.15 mass% silicon
■   2   mass% magnesium + 0.15 mass% silicon
□   1   mass% magnesium + 0.15 mass% silicon
○   0.5   mass% magnesium + 0.15 mass% silicon

the metal could be described unequivocally by standard physical properties appropriate to the context,[5] e.g. molality or mole fraction for a solution and atoms and conservation of mass in for heterogeneous systems. However, in real metal products the quantity and distribution of hydrogen among various occluded forms is indeterminate, variable and subject to the influences of extraneous factors introduced in the course of manufacturing operations. Hence any measure of the occluded hydrogen must be independent of the state of occlusion and for this purpose, an empirical quantity, the hydrogen content, is defined as follows.

4.4.2 HYDROGEN CONTENT

The hydrogen content is the total quantity of hydrogen present in a specified mass of metal, irrespective of its distribution among occluded forms, i.e. solution and traps of various kinds. It is specified as the volume of diatomic hydrogen gas, V/cm^3, measured at 273 K and 101 kPa that can be extracted from 100 g of the metal, abbreviated to "cm^3/100 g". Familiarity in use has conferred the status of a universal industry standard on this particular definition for the aluminium-hydrogen system and other gas-metal systems. It is used to quantify occluded hydrogen in the following chapters concerned with technical matters. Its applications include setting limits for quality control, assessing the efficiency of degassing operations and investigating manufacturing operations yielding defective products.

The measure of hydrogen content as the volume of a hypothetical equivalent gas phase[28-33] is now firmly established, although with hindsight a direct measure in terms of the mass of hydrogen *in situ* in the metal might have been more logical.

4.5 REFERENCES

1. A. Taylor, *An Introduction to X-ray Metallography*, Chapman & Hall, London, 1952.
2. W. Hume-Rothery, *The Structure of Metals and Alloys*, Institute of Materials, London, 1950.
3. L. Pauling, *The Nature of the Chemical Bond*, Oxford University Press, London, 1960, 23 & 32.
4. C.A. Coulson, *Valence*, Clarendon Press, Oxford, 1952, 19 & 119.
5. M.L. McGlashan, *Physicochemical Quantities and Units*, Monographs for Teachers No. 15, The Royal Institute of Chemistry, London, 1971, 72 & 101.
6. S. Glasstone, *Thermodynamics for Chemists*, Van Nostrand, New York, 1947, 350–355.
7. G.N. Lewis, M. Randall, K.S. Pitzer and L. Brewer, *Thermodynamics*, McGraw-Hill, New York, 1961, 243–249.
8. A. Sieverts, *Z. Metallkunde*, **21**, 1929, 37–45.
9. J.C. Slater, *Quantum Theory of Matter*, McGraw-Hill, New York, 1968.
10. M.J. Stott and E. Zaremba, *Phys. Rev. B*, **22**, 1980, 1564.
11. M.J. Stott and E. Zaremba, *Solid State Commun.*, **32**, 1979, 1297.
12. J.K. Nørskov and N.D. Lang, *Phys. Rev. B*, **21**, 1980, 2136.

13. J.K. Nørskov, *Phys. Rev. B*, **26**, 1982, 2875.
14. J.K. Nørskov, *Phys. Rev. Lett.*, **48**, 1982, 1620.
15. J.K. Nørskov, F. Besenbacher, J. Bøttiger, B.B. Nielson and A.A. Pizarev, *Phys. Rev. Lett.*, **49**, 1982, 1420.
16. M.J. Puska, R.M. Niemenin and M. Manninen, *Phys. Rev. B*, **24**, 1981, 3037.
17. M.S. Daw and M.I. Baskes, *Phys. Rev. B*, **29**, 1984, 6443.
18. M.S. Daw and M.I. Baskes, *Phys. Rev. Lett.*, **50**, 1983, 1285.
19. F. Herman and S. Skillman, *Atomic Structure Calculations*, Prentice-Hall, New Jersey, 1963.
20. R.M. Barrer, *Disc. Faraday Soc.*, **4**, 1948, 68.
21. *International Critical Tables*, **7**, 231.
22. S. Glasstone, *Thermodynamics for Chemists*, Van Nostrand, New York, 1947, 310–311.
23. G.N. Lewis, M. Randall, K.S. Pitzer and L. Brewer, *Thermodynamics*, McGraw-Hill, New York, 1961, 419–428.
24. R.H. Fowler and C.J. Smithells, *Proc. Roy. Soc.*, **160**, 1937, 37.
25. S. Glasstone, *Thermodynamics for Chemists*, Van Nostrand, New York, 1947, 151 & 199.
26. G.N. Lewis, M. Randall, K.S. Pitzer and L. Brewer, *Thermodynamics*, McGraw-Hill, New York, 1961, 226.
27. R.A. Swalin, *Thermodynamics of Solids*, John Wiley & Sons, New York, 1962, 36–38.
28. H. Schenck and K.W. Lange, *Arch. Eisenhutt. Wes.*, **37**, 1966, 739.
29. L. Luckmeyer-Hasse and H. Schenck, *Arch. Eisenhutt. Wes.*, **6**, 1932–33, 209.
30. A. Sieverts and H. Hagen, *Z. phys. Chem.*, **169**, 1934, 237.
31. W. Eichenauer and A. Pebler, *Z. Metallkunde*, **48**, 1957, 373.
32. P. Rontgen and F. Moller, *Metallwirt., Metallwiss., Metalltech.*, **13**, 1934, 81 & 97.
33. A. Sieverts, *Ber. dt. Chem. Ges.*, **45**, 1912, 221.
34. C.E. Ransley and H. Neufeld, *J. Inst. Metals*, **74**, 1948, 599.
35. D.E.J. Talbot, *International Met. Reviews*, **20**, 1975, 166.
36. D.E.J. Talbot and D.A. Granger, *J. Inst. Metals*, **92**, 1963–4, 290.
37. C. Renon and J. Calvet, *Mem. Sci. Rev. Met.*, **58**, 1961, 835.
38. C. Renon and J. Calvet, *Mem. Sci. Rev. Met.*, **60**, 1963, 620.
39. A.G. Bondarenko and A.B. Khmelinen, *Phys. Stat. Solid*, **88**, 1985, 121.
40. R.A. Swalin, *Thermodynamics of Solids*, John Wiley & Sons, New York, 1962.
41. D. Mclean, *Grain Boundaries in Metals*, Oxford University Press, London, 1957.
42. W.A. Harrison, *Pseudopotentials in the Theory of Metals*, Benjamin, New York, 1966.
43. M. Rasolt and R. Taylor, *Phys. Rev. B*, **11**, 1975, 2717.
44. M. Manninen, P. Jena, R.M. Nieminen and J.K. Lee, *Phys. Rev. B.*, **24**, 1981, 7057.
45. M.I. Baskes and C.F. Melius, *Phys. Rev. B.*, **20**, 1979, 3197.
46. R.A. Johnson and W.D. Wilson, *Interatomic Potentials and Simulation of Lattice Defects*, P.C. Gehlen, J.R. Beeler and R.I. Jaffee, eds., Plenum Press, New York, 1971.

47. W.D. Wilson, *Phys. Rev. B,* **24**, 1981, 5616.

48. W. Kohn and L.J. Sham, *Phys. Rev.*, **140**, 1965, A1133.

49. S.M. Myers, S.M. Picraux and R.E. Stoltz, *J. Appl. Phys.*, **50**, 1981, 5710.

50. D.M. Henry, *"The Nature and Effects of Hydrogen in Weldalite Aerospace Alloy and Other Aluminium-Lithium Alloys"*, Ph.D. Thesis, Brunel University, 1995.

51. R.C. Dickenson, K.R. Lawless and K. Wefers, *Scripta Metallurgica,* **22**, 1988, 917.

52. R. Balasubramanium, D.J. Duquette and K. Rajan, *Acta Metall. Mater.*, **39**, 1991, 2607.

53. C.E. Ransley and D.E.J. Talbot, *J. Inst. Metals*, **88**, 1959–60, 150.

54. A.J. McAlister, *Bulletin of Alloy Phase Diagrams*, **3**, 1982, 177.

55. J.H.E. Jeffes and H. McKerell, *J. Iron Steel Inst.*, **202**, 1964, 666.

56. E. Veleckis, *J. Less-Common Metals,* **73**, 1980, 49.

57. S. Aronson and F.J. Salzano, *Inorg. Chem.*, **8**, 1969, 1541.

58. T.G. Pearson, *The Chemical Background of the Aluminium Industry*, The Royal Institute of Chemistry, London, Monograph No.3, 1955, 46.

59. M. Hansen and K. Anderko, *Constitution of Binary Alloys*, McGraw-Hill, New York, (3), 1958.

60. F. Keller and C.M. Craighead, *Trans. AIME*, **122**(4), 1936, 315–320.

61. C.E. Ransley and D.E.J. Talbot, *J. Inst. Metals.*, **84**, 1955–6, 445.

5. The Determination of Hydrogen in Aluminium and its Alloys

Two techniques, vacuum extraction[1] and the Telegas[2] are the dominant accurate methods for determining the hydrogen content of solid and liquid aluminium and its alloys. They are supplemented but not superseded by other methods which offer some practical advantages but which are less precise and reliable.

The following review summarises the principles underlying available analytical methods, describes their application, and assesses current practices and developments.

5.1 GENERAL PRINCIPLES

Analytical methods are based on either collection of the gas extracted from suitable samples or measurement of the activity of hydrogen in solution.

5.1.1 COLLECTION METHODS

Collection methods are applicable to both solid metal and to samples taken from liquid metal quenched to retain the hydrogen in solution. The measurements are inherently difficult because the quantities of hydrogen recovered from samples are small and easily confused with spurious hydrogen generated from reaction of the metal samples with traces of water on their surfaces and with water vapour released from the walls of the equipment used.

The classic method is vacuum extraction in which a solid sample of the metal is heated in a continuously evacuated system and the gas recovered is collected for measurement and analysis.[1] It is slow and requires meticulous attention but it is the most reliable method.

For convenience in routine use, Degrève[3] devised an alternative approach based on the evolution of hydrogen from a melted sample into a metered flowing carrier gas that is continuously analysed for hydrogen. The advantages over vacuum extraction are speed of operation and reduced dependency on operator skill but they are offset by lower reliability and restrictions in the range of alloys for which the method is suitable.

5.1.2 METHODS BASED ON MEASUREMENTS OF HYDROGEN ACTIVITY

Standard methods for determining the activity of a species in solution include:
1. Measurement of the pressure of a gas phase in equilibrium with the solution.
2. Measurement of equilibrium electrochemical potentials for dissolution.

The measurement of pressure in the gas phase is particularly suitable for hydrogen dissolved in liquid aluminium because equilibrium can be established in a reasonable time. The gas is virtually ideal and the solution exhibits Henrian activity at the temperatures and pressures of interest, so that the quantity in solution is given directly by Sieverts' isotherm, derived as Eqn 4.6 in Chapter 4:

$$m/m^{\ominus} = K^{1/2} \cdot (p/p^{\ominus})^{1/2} \tag{5.1}$$

The Telegas instrument[2] exploits the principle to give rapid direct readings for liquid metal with a reliability approaching that of the vacuum extraction method.

Empirical tests sometimes used in foundries to indicate excessive hydrogen contents are based on the nucleation of bubbles in samples taken from liquid metal placed under reduced external pressure. They are rough guides to the hydrogen activity. Examples are the Straube Pfeiffer vacuum solidification test[4] and Dardel's initial bubble test.[5]

The measurement of electrochemical potentials is an attractive principle in view of its successful application in oxygen probes for liquid steel[6] but difficulties are associated with the low-temperature solid electrolytes required and the technique does not seem to offer significant advantages over the simpler Telegas method.

5.2 THE VACUUM EXTRACTION TECHNIQUE

The vacuum extraction technique,[1, 7] is the accepted standard reference method for hydrogen content determinations; it is generally known as Ransley's method, recognising the originator. The equipment is versatile and with minor modifications it has also been applied to determine the solubilities of hydrogen in liquid and solid aluminium and its alloys by Sievert's method and some corresponding diffusivities, as described in Chapters 6 and 7.

5.2.1 OPERATING PRINCIPLES

Hydrogen is extracted from a sample heated in an evacuated vessel to a temperature close to but below the solidus temperature of the metal. The gas is continuously pumped to a calibrated low-pressure system for collection and measurement. The use of solid samples confers several benefits:

1. Complications in providing a containing crucible are avoided.
2. The generation of spurious hydrogen by reaction of volatile elements with the walls of the equipment can be controlled.
3. Corrections for spurious hydrogen generated by reaction of traces of water vapour with the hot sample surface are lower than for fusion extraction and are more consistent.

The speed of the determination depends on the diffusion-controlled desorption of the gas from the sample. As explained in Chapter 4, the hydrogen in manufactured aluminium products is distributed between solution and micropores and so it diffuses through the sample in a field of traps. This reduces its mobility by an order of magnitude below that expected from the true diffusion coefficients, D_{true}, which do not normally apply to manufactured products but only to laboratory samples consolidated by heat-treatment in vacuum and then loaded with hydrogen to ensure that it is wholly in solution.[9, 10] Nevertheless, hydrogen evolved from small samples of manufactured metal over periods of a few hours at constant temperature can be fitted to standard solutions of Fick's equation,[11] yielding empirical pseudo diffusion coefficients, $D_{apparent}$. The value of the pseudo coefficient for pure aluminium as a function of temperature is:[12]

$$D_{apparent} = 1.2 \times 10^5 \exp \frac{-16900}{T/K} \ cm^2 \ s^{-1} \qquad (5.2)$$

It is insensitive to the source of the metal, and was mistaken for the true diffusion coefficient, D_{true}, before the nature of hydrogen occlusion in manufactured aluminium products was appreciated. Using $D_{apparent}$ in place of D_{true} in solutions to Fick's equation yields reasonable approximations for the movement of hydrogen in manufactured products. In particular, it can be used to predict the time required for the extraction of hydrogen from a sample to any prescribed degree of completion and to process data on line by computer, if required, as described in Section 5.2.4.

5.2.2 EQUIPMENT

5.2.2.1 Gas Collection and Analysis System

The equipment, illustrated schematically in Figure 5.1 is constructed of hard glass with a graded seal to a lead-glass bulb where various direct joints are made to platinum.

The demountable spherical O-ring joint, B, receives a sample tube, A, made of 18 mm bore clear fused silica in which a metal sample, C is heated by a radio frequency (RF) induction coil, D. It affords access, via a 20 mm bore tap, E, and a liquid-nitrogen cooled trap, F, to a two-stage mercury-diffusion collection pump, G, which delivers the gas extracted from the sample into a system with a calibrated volume. The size of the vessel, H, is chosen to adjust the total volume of the system to about 1000 cm³ so that the gas collected from a standardised 8–10 g sample is contained at pressures in the optimum range, 0–0.1 Pa, of a McLeod gauge, J. The facilities for analysis are a platinum filament and a palladium tube. The filament, K, is a loop of 0.1 mm diameter platinum wire carried

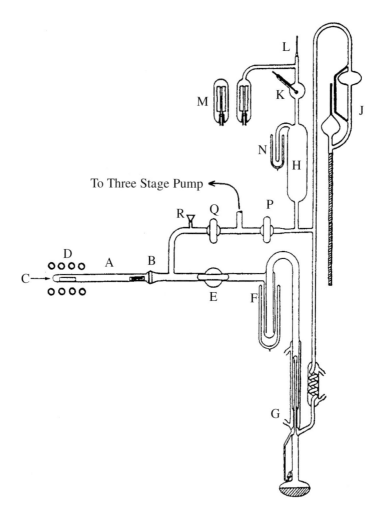

Fig. 5.1 Schematic diagram of vacuum extraction equipment for hydrogen content determinations. (Ransley and Talbot)[1]

A.	Sample Tube	J.	McLeod Gauge
B.	Demountable O-ring Joint	K.	Platinum Filament
C.	Sample	L.	Palladium Tube
D.	RF Induction Coil	M.	Pirani Gauge
E.	Large Bore Tap	N.	Liquid Nitrogen Cooled Tube
F.	Liquid Nitrogen Cooled Trap	P.	Tap to Mercury Diffusion Pump
G.	Collection Pump	Q.	By-Pass Tap
H.	Dead Space Vessel	R.	Valve to Relieve Vacuum

Fig. 5.2 General view of vacuum extraction equipment for hydrogen content determinations.

on platinum leads. The palladium tube, L, is 50 mm long × 1.3 mm diameter with 0.15 mm wall thickness. It is closed at one end and at the open end it is gold-soldered into a platinum thimble sealed to lead-glass. A small electric heater is provided to heat the tube to 700°C. The system includes a sensitive Pirani gauge, M, calibrated against the McLeod gauge, J. Liquid nitrogen can be applied locally to the side-tube, N, to condense spurious water vapour without significantly affecting the temperature of the permanent gases present.

The system is evacuated by a 3 stage mercury diffusion pump, not shown in Figure 5.1, backed by a rotary pump and can be isolated from it by a tap, P. A by-pass tap, Q, allows evacuation of the sample tube after loading a sample, without admitting air to the system.

The general construction is shown in Figure 5.2. The glasswork is erected on a framework permitting easy access for glass-blowing and servicing. The diffusion pumps are electrically heated. A furnace to pre-bake the sample tube is suspended from an arm pivoted on the framework, facilitating its alignment with and movement over the end of the sample tube. Its temperature is controlled by a thermocouple at its mid point.

The specialised calibration procedures for the McLeod gauge and the volume of the collection system are described in Appendix 1.

5.2.2.2 Arrangements for Heating Samples

Radio frequency (RF) induction is the best option[7] because it heats the sample directly, conferring the following benefits:

1. The temperature of the sample is raised rapidly to any prescribed constant value; for example, a 1 kw heater at full power can raise the temperature of a 10 mm diameter sample of pure aluminium, 40 mm long to 500°C in ~60 s and maintain it at that temperature with ~8% power. Hence the desorption of hydrogen is almost isothermal, shortening the time to complete a determination and providing an option for on-line computer data processing.
2. The sample tube remains cool, so that any volatile metals, magnesium, sodium, lithium or zinc evaporated from the sample during the extraction of hydrogen condense on the walls of the tube close to the sample. Hence a simple straight tube as illustrated in Figure 5.3a is suitable.

To avoid the inconvenience of routinely sealing thermocouples through the walls of the vacuum system to monitor and control the sample temperature, the RF heater is timed to supply full power to raise the sample temperature to the prescribed value and switch to reduced power to maintain it, using programmes predetermined with dummy samples fitted with internal thermocouples. The programme settings are sensitive to alloy compositions but not to the small dimensional errors expected in routine sample preparation.

If RF induction heating is not available, the sample can be heated by an electric resistance furnace placed over the end of the sample tube[1] but this incurs the following disadvantages:

1. The determination is prolonged by about an hour because of the delay in heating the sample due to poor thermal coupling through the vacuum.
2. The vapours of volatile metals evaporated from the sample cannot condense on the heated walls of the tube and are directed along its axis so that the condensate trap illustrated in Figure 5.3b is needed to avoid generation of spurious hydrogen, as explained in Section 5.2.3.2.

5.2.3 PROCEDURES

5.2.3.1 Sample Preparation

For bulk materials, it is most convenient to use fine-turned cylindrical samples, 40 × 10 mm in diameter. They are prepared to a reproducible, scrupulously clean, fine surface finish. The best method is to machine them dry, using well-maintained round-nosed tools with high top and side rakes, The finishing cut is light and fast with a slow feed. It is vitally important to remove all of the original surface, which is inevitably

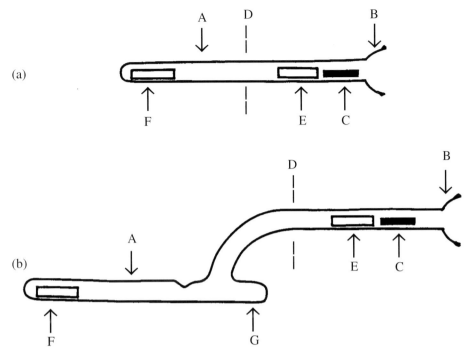

Fig. 5.3 Sample tubes for vacuum extraction (a) for use with RF induction heating and (b) for use with external furnace heating.

A.	18 mm Bore Clear Silica Tube	
B.	Demountable O-Ring Joint	
C.	Manipulator	
D.	Limit of Baking	

E.   Sample During Pre-Baking
F.   Sample During Extraction
G.   Condensate Trap

contaminated. The machined samples are washed in dry high-purity benzene, dried and used immediately.

Samples cut from thick sheet material can be milled on both surfaces and the edges filed clean in hygienic conditions. Materials such as thin sheet, wire and foil which are difficult to prepare mechanically are cut to appropriate sizes and etched in 10% caustic soda to remove the original surface, rinsed in dilute nitric acid and washed successively in distilled water, alcohol and acetone. Preparation by abrasion with grinding papers produces a surface source of spurious hydrogen and is best avoided.

5.2.3.2 Conditioning the Sample Tube

The surface of silica invariably absorbs traces of water from the normal atmosphere even if only momentarily exposed. The absorbed water is a serious source of spurious hydrogen by reaction with the hot sample and any condensate, so that it must be driven off by pre baking the tube, including the condensate trap illustrated in Figure 5.3b if the sample is to be heated by an external furnace. The sample is placed in the sample tube near the open end together with a small rod, C in Figure 5.3, of a magnetic material e.g. ferritic stainless steel to manipulate it using an external permanent magnet. The tube is assembled on the apparatus by the demountable joint and evacuated through the by-pass tap, Q in Figure 5.1. The whole of the tube is pre-baked at 800°C for 1 hour, except for the region around the sample and the O ring joint, B. When the tube has cooled, the integrity of the system is tested by closing taps P and Q in Figure 1 with tap E open. The criterion to proceed is that the gas accumulated in 15 minutes is negligible. If the test is satisfactory, tap P is reopened momentarily to restore the vacuum and closed again. Liquid nitrogen is applied to traps F and N in Figure 5.1 and losses by evaporation are replenished throughout the determination to eliminate residual traces of water vapour, which otherwise generate spurious hydrogen.

5.2.3.3 Extraction of Hydrogen

The sample is pushed to the extraction position, shown in Figure 5.3, by the manipulator, using an external magnet to move and return it to its original position. The RF coil or external furnace as appropriate is placed over the sample and power is applied.

The pressure in the collection system is monitored at intervals by the McLeod and Pirani gauges until it is judged to be virtually constant, indicating the end-point. Figure 5.4 gives examples of hydrogen evolution curves recorded in this way.

5.2.3.4 Correction for Residual Gases

The gas collected is analysed for hydrogen at constant volume. The platinum filament is heated to 700°C to test its thermal stability and if, as is usual, there is no change in pressure the palladium tube is heated to 700°C for 10 minutes, allowing the hydrogen to diffuse out. The residual gas is usually < 5% of the quantity collected. It is a mixture of carbon monoxide, methane and nitrogen. Its pressure is recorded to apply as a correction.

5.2.3.5 Correction for Extraneous Hydrogen

The volume of gas collected must also be corrected for extraneous hydrogen associated with the sample surface area, generally known as the *surface correction*[1] It can be evaluated from blank determinations on representative degassed samples after re-preparation to restore the original surface condition. The correction is determined by the alloy composition, the surface area of the sample and the duration of vacuum extraction. Hence values for the correction must be obtained from blank determinations of the same duration as the hydrogen content determinations to which they are applied. Standardised corrections are in general use and are adequate provided that the hydrogen contents measured are not too small, e.g. < 0.15 cm³/100 g. Typical values quoted for the volume of extraneous hydrogen produced per cm² of sample surface area are 2.0×10^{-4} cm³ for pure aluminium and 3.5×10^{-4} cm³ for AA 2024 alloy,[1] yielding corrections of the order of 10 to 15% of the hydrogen collected from the standard cylindrical samples 10 mm diameter × 40 mm.

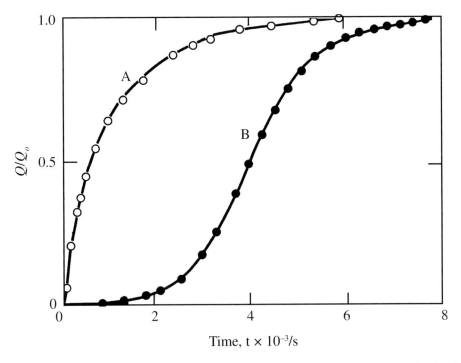

Fig. 5.4 Evolution of hydrogen from 10 mm diameter cylindrical sample of a duralumin-type alloy (AA 2024) during vacuum extraction at 500°C. Fraction collected, Q/Q_o, as a function of time (Ransley and Talbot).[1] Curve A: Sample heated by radio-frequency induction and Curve B: Sample heated by an external furnace.

5.2.3.6 Care and Maintenance of Sample Tubes

Fused silica is the only inexpensive material fulfilling all of the needs for sample tubes:
1. Easy fabrication and availability of standard O-ring joints made of the same material.
2. Resistance to high temperatures and thermal shock.
3. Transparency to facilitate sample manipulation and inspection for condition.
4. Chemical inertness to air and cleaning agents.

The tubes must be kept scrupulously clean and dry. When a tube is dismantled at the end of a determination, there is usually a condensate of volatile metals distilled from the sample surface. Even commercial grades of pure aluminium can yield a thin deposit if they contain traces of sodium or magnesium. Alloys with appreciable magnesium contents, e.g. in the AA 2000, 5000 and 6000 series produce bright, reflective condensates. The condensates must be dissolved in a dilute acid and the tube rinsed with distilled water and dried before re-use. It is good practice to store the dried tubes in a laboratory oven controlled at ~150°C. With repeated use, silica at the outside walls of the tube recrystallises, but despite the outward appearance, close inspection reveals that the inside walls remain in good condition.

Alloys containing high magnesium contents, e.g. AA 5182 and AA 7075 and/or lithium contents, e.g. AA 8090 are vulnerable to spurious results if the samples are heated by an external furnace. The cause is severe attack by lithium and/or magnesium vapours on the hot walls of the sample tube, exposing and reacting with hydration deeper in the silica structure, yielding hydrogen. The attack is manifest as a permanent brown discoloration and affected tubes are irrecoverable. The problem is not serious if the samples are heated by RF induction so that the silica remains cool and the tubes are well maintained and prebaked. The reliability of determinations on aluminium–lithium alloys must not be confused with the very slow desorption of hydrogen characteristic of these alloys, as explained in Chapter 7.

5.2.4 ON-LINE DATA PROCESSING BY COMPUTER

Wood et al.[7] applied on-line data processing, based on experience that the extraneous hydrogen associated with the surface of a sample held at constant temperature accumulates at a constant rate. Figure 5.5 gives an example for hydrogen collected during a determination in which a cylindrical sample is heated isothermally by RF induction. Curve A, which records the determination, approaches an asymptote of positive slope showing that hydrogen is collected at a constant rate long after the sample has yielded all of its hydrogen content. The line B records a subsequent blank determination on the sample left undisturbed in the evacuated sample tube after the actual determination; it confirms the collection of hydrogen at the same constant rate. The slope of the asymptote varies between one determination and another, even for replicate samples, suggesting that the surface correction is influenced by a variation in the condition of the equipment specific to every determination, e. g. variation in the degree to which a low water vapour potential can persist in the surface structure of the silica tube despite prebaking, as a precursor to hydrogen formed by reaction at the hot metal surface.

The data processing applies a simple mathematical model subtracting the surface hydrogen and fitting an appropriate solution of Fick's equation[11] to the remainder which is desorbed from the sample under diffusion control, as justified in Section 5.2.1.

For a cylinder of aspect ratio > 4, radius, r, with a hydrogen content, Q_o, initially distributed as a uniform concentration, c_o, and subject to the initial and boundary conditions:

$$c = c_o \text{ for } 0 < r < r, t = 0 \quad \text{and} \quad c = 0 \text{ for } r = r, t > 0$$

the hydrogen fraction, Q/Q_o, desorbed in time, t, from homogeneous solution has the form:[11]

$$Q/Q_o = A[1 - \exp(-Bt)] \qquad (5.3)$$

Incorporating a small zero offset, Δt to correct for the short initial period (~60 s) of non isothermal operation while the sample is heated to the extraction temperature and

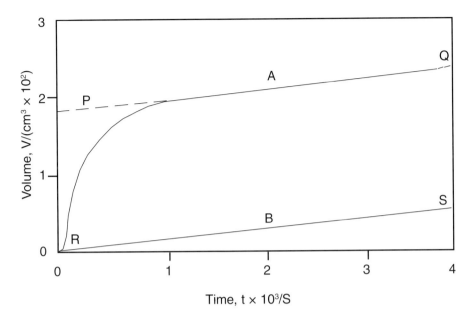

Fig. 5.5 Volumes of hydrogen collected during vacuum extraction of a 10 mm diameter × 40 mm cylindrical sample of AA 2024 alloy at 500°C. Sample heated isothermally by radio-frequency induction. (A) Hydrogen content determination. The volume collected is recorded as a curve of diminishing slope approaching the asymptote PQ, showing continuing evolution of hydrogen generated at the sample surface and (B) Subsequent blank determination on the sample left undisturbed in the evacuated sample tube after the real determination. The collection of surface hydrogen continues at the same rate, as shown by the linearity of the record, RS, with the same slope as PQ.

inserting the linear term, $C(t - \Delta t)$, to allow for the accumulation of the surface hydrogen the total quantity of gas, Q_{TOTAL}, collected as a function of time is:

$$Q_{TOTAL} = Q/Q_o + C(t - \Delta t) = A[1 - \exp\{-B(t - \Delta t)\}] + C(t - \Delta t) \qquad (5.4a)$$

The input for data processing is the output potential, E, of the Pirani gauge, which is proportional to Q_{TOTAL}, so that Eqn 5.4a is rewritten in terms of E, modifying the values of the constants A, B and C to include the Pirani gauge calibration factor:

$$E = A[1 - \exp\{-B(t - \Delta t)\}] + C(t - \Delta t) \qquad (5.4b)$$

The Pirani signal is sampled at frequent intervals, yielding actual data points, $(E, t)_{ACTUAL}$ that are applied through an interface to a computer programmed with a curve-fitting algorithm calculating the constants, A, B and C in Eqn 5.4b and using them to predict future data points, $(E, t)_{PREDICTED}$. The reliability of the calculations is upgraded progressively as more data points become available. When sufficient points have been accumulated to ensure that the fit between predicted and actual values of the data points is within predetermined

Table 5.1 Data Processing Record for a Hydrogen Content Determination by Vacuum Extraction of a 10 mm diameter × 40 mm cylindrical sample at 500°C

Time t/s	Actual Pirani Gauge Reading E/mV	Constants in Equation 5.4b			Predicated Pirani Gauge Reading* E/mV	Standard Deviation[†]	Predicted Final Result $cm^3/100$ g
		A	B	C			
120	11.5	–	–	–	–	–	–
240	33.2	–	–	–	–	–	–
360	42.5	–	–	–	–	–	–
480	46.8	–	–	–	–	–	–
600	48.5	–	–	–	–	–	–
720	50.2	–	–	–	–	–	–
840	51.0	–	–	–	–	–	–
960	51.9	49.9	0.11	0.36	51.5	0.83	0.19
1080	53.2	48.6	0.23	0.37	52.5	0.80	0.18
1200	53.9	47.9	0.30	0.38	53.5	0.77	0.18
1320	54.6	47.6	0.32	0.38	54.3	0.73	0.18
1440	55.5	47.1	0.36	0.39	55.3	0.72	0.18
1560	56.6	46.9	0.37	0.39	56.1	0.69	0.18
1680	57.2	46.7	0.38	0.39	57.0	0.67	0.18
1800	57.7	46.8	0.38	0.39	57.8	0.64	0.18
1920	58.7	46.6	0.39	0.39	58.6	0.62	0.18
2040	59.6	46.6	0.39	0.39	59.5	0.60	0.18
2160	60.2	46.6	0.39	0.39	60.3	0.59	0.18
2280	61.3.	46.5	0.40	0.40	61.1	0.57	0.18
2400	62.3	46.5	0.40	0.40	62.0	0.56	0.18

* Calculated from values of the constants in Colums 3–5, derived by fitting Eqn 5.4b to the accumulated actual data points (E, t), given in Columns 1 and 2.

† For pairs of actual and predicted readings to date.

statistical limits, the value of E for $t = \infty$ is predicted and used with the Pirani gauge calibration factor and the known volume of the gas collection system to calculate the hydrogen content of the sample. Table 5.1 illustrates the data processing operation. Every line in Columns 3 to 5 gives progressively updated values of the constants A, B and C calculated from the past series of actual readings given in Column 2. They are used to predict the next reading on the line below in Column 6. The sequence of standard deviations given in Column 7 show a progressively closer fit between the actual and calculated data points. Column 8 gives corresponding predictions of the hydrogen content. Thus the following advantages are secured:

1. The end-point is assessed impartially and the result predicted to prescribed accuracy.
2. The correction for extraneous hydrogen is specific to the determination in progress.

5.2.5 VALIDITY OF RESULTS

The accuracy of the method depends mainly on reliable corrections for spurious gases. Unless they are low and reproducible, the standard of work is inadequate. Abnormal early gas evolution suggests contamination of the sample and indefinite end-points suggest contaminated equipment. The condensates of active elements, e.g. magnesium in pre-baked sample tubes yield no hydrogen if redistilled[1], showing that no hydrogen is lost by gettering.

5.2.5.1 Reproducibility of the Method

Ransley and Talbot[1] illustrated the reproducibility by the following replicate results:

1. 14 replicate determinations on an extruded 99.8% pure aluminium rod yielded results in the range 0.24 to 0.28 cm³/100 g, with a standard deviation for a single result of 0.014 cm³/100 g.
2. Duplicate determinations on 43 samples of commercial grades of pure metal and alloys with < 1.0% magnesium yielded results with a standard deviation for a single result of 0.015 cm³/100 g.
3. Duplicate determinations on 16 samples of alloys with 5% magnesium and hydrogen contents in the range 0.14 to 0.97 cm³/100 g, yielded results with a standard deviation for a single result of 0.023 cm³/100 g.

5.2.5.2 Absolute Accuracy

Figure 5.10 gives an example of the good agreement between vacuum extraction results and corresponding results obtained with the Telegas instrument, which is based on independent principles explained in Section 5.3.

5.3 THE TELEGAS

The Telegas is a rapid direct reading instrument measuring the hydrogen content of liquid metal, designed and developed by Ransley et al.[2] It fulfils two functions:

1. In research and development on liquid metal, it can replace time-consuming determinations by the vacuum extraction technique.
2. In production, it can monitor the hydrogen content of liquid metal for quality control.

The principle is to determine the activity of diatomic hydrogen in equilibrium with the solution. Since hydrogen is virtually ideal, this can be accomplished by measuring the hydrogen pressure established in a gas-filled cavity created within the liquid metal. The hydrogen content is then given by Sieverts' isotherm, Eqn 5.1, usually simplified to:

$$G = S\left(\frac{p}{p^{\ominus}}\right)^{\frac{1}{2}} \tag{5.5}$$

where G is the hydrogen content, p is the measured pressure and S is the solubility for the pressure $p^{\ominus}$.

Hydrogen withdrawn from solution in making the measurement reduces the concentration locally and this imposes a constraint on the permissible volume of the measuring system, illustrated by the following simple calculation. A typical requirement is to measure a hydrogen content of 0.3 cm³/100 g in liquid pure aluminium at 700°C, for which $S = 0.9$ cm³/100 g for $p^{\ominus} = 10^5$ Pa (1 atm.),[13] as explained in Chapter 6. If the hydrogen is collected in a system with a total volume equivalent to 1 cm³ at ambient temperature (20°C) and the error is to be restricted to, say, 5%, the hydrogen supplied to the measuring system must be drawn from a catchment volume of the liquid metal surrounding the probe corresponding to a sphere of at least 4 cm in diameter. In quiescent liquid metal, the transport of hydrogen to the sensor is diffusion-controlled, so that for rapid response the measuring system must be very small to minimise the corresponding catchment volume.

5.3.1 VALIDATION OF PRINCIPLE

To validate the principle, Ransley et al.[2] constructed a simple micro-system in which the equilibrium hydrogen pressure was sampled by immersing into the liquid metal a probe comprising a small palladium tube enclosed in a thin close-fitting porous alumina sheath. The pressure was transmitted by small-bore stainless steel tubing to an evacuated mercury manometer made of glass capillary tubing. The volume of the system was 0.6 cm³. Many tests in laboratory melts of pure aluminium were made with this equipment. The recorded pressure rose steadily, reaching a maximum after 8–10 minutes when the liquid metal temperature was recorded and samples of the metal were cast in chilled moulds. The hydrogen contents are calculated by application of Eqn 5.5, inserting critically assessed values for solubility, as given in Chapter 6.[13] For comparison, the hydrogen contents of the chilled metal samples were determined by the vacuum extraction technique. Results obtained for pure aluminium are compared in Figure 5.6. In view of the independent principles of the methods by which the results were obtained, there is good qualitative agreement, but the results from the pressure measurements were systematically lower than those from vacuum extraction. This reflects two inherent weakness in the arrangement which delay the approach to equilibrium:
1.   The system is passive and the probe collects hydrogen from a static catchment volume.

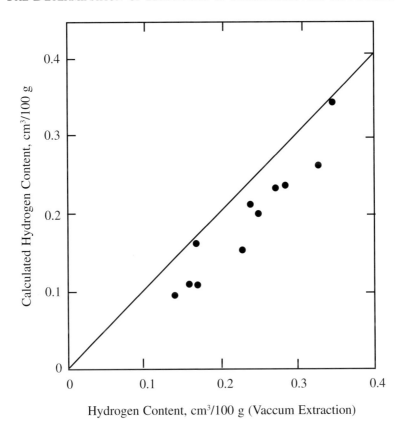

Fig. 5.6 Hydrogen contents of liquid 99.2% pure aluminium calculated from equilibrium hydrogen pressures established in a palladium tube protected from the metal. Comparison with hydrogen contents of corresponding cast samples determined by vacuum extraction. (Ransley *et al.*).[2]

2. Transfer of hydrogen across the metal/gas interface is impeded by oxide films formed from oxygen in the air carried in the porous alumina sheath.

    Attempts to eliminate these impediments, including argon flushing the alumina sheath and varying the probe geometry were unsuccessful and were abandoned in favour of designing an active system.

5.3.2 MICRO CARRIER GAS SYSTEM

To overcome the passivity problems, an active micro carrier gas circuit was developed in which a small mass of an inert carrier gas is recirculated in a system embodying a probe in which the carrier gas experiences vigorous contact with the liquid metal, and a device

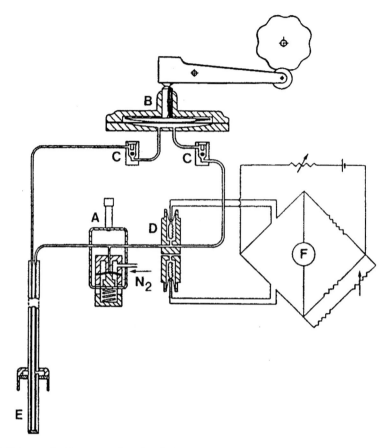

Fig. 5.7 Schematic diagram of recirculating carrier gas micro-system, the "Telegas", for determining hydrogen contents in liquid aluminium and aluminium alloys. (Ransley *et al*.).[2]

A.	Nitrogen Admission Valve	D.	Katharometer
B.	Recirculating Pump	E.	Probe
C.	Recirculating Pump Valves	F.	Bridge Circuit Millivoltmeter

recording the accumulation of hydrogen in the gas. For economy and design simplicity, dry oxygen free nitrogen is selected as the carrier gas.

The circuit is illustrated diagrammatically in Figure 5.7. It comprises a carrier gas admission valve, A, a mechanical circulating pump, B, with associated valves, C, a Katharometer[14] (a hot-wire conductivity gauge), D, and a probe, E, which is immersed in the liquid metal. Gas connections are made with 1 mm bore copper or stainless steel tubing as appropriate.

Every component is designed and specially constructed to contribute the minimum practicable volume to the system as follows:

	Actual Volume/cm³	**Approximate Nominal Gas Volume at 0°C/cm³**
Katharometer	0.4 (at 20°C)	0.4
Circulating Pump	0.5 (at 20°C)	0.5
Valves	0.1 (at 20°C)	0.1
Connecting Tubing	0.7 (at 20°C)	0.7
Probe	4.0 (at 700°C)	1.1
Total		2.8

With such a small mass of circulating gas, the function of the instrument is vulnerable to small actual or virtual leaks, especially of water vapour which produces spurious hydrogen on contact with the hot metal. Actual leaks are avoided by skilled construction. Virtual leaks are avoided by minimal use of absorbent materials, e.g. rubber, in the design of pumps and valves and by maintaining scrupulous internal cleanliness. The essential components are:

Circulating Pump and Associated Valves
The requirements are a high ratio of swept to total volume and sufficient output pressure to overcome resistance in the system and the small hydrostatic head of metal in the probe. In the design evolved, a thin metal diaphragm is elastically deformed into a shallow depression by a cam-operated ram. Inertial metal disc micro-valves are fitted.

Carrier Gas Admission Valve
The admission valve comprises a spring-loaded neoprene pad closing a 1 mm diameter orifice connected by a short capillary tube to the gas circuit on the upstream side of the pump.

Katharometer
The katharometer is specially constructed to have the minimum volume. Two identical cells, each with a volume of 0.4 cm³ are bored in a copper block. Each cell contains a spiral platinum filament, 25 μm in diameter with a resistance of 15 Ω at ambient temperature, and communicates by a short channel to separate capillary gas lines. One cell is selected as the active cell and its gas line forms part of the nitrogen carrier gas circuit; the other is the comparator cell and its gas line is open to air to serve as the comparator gas. The filaments are connected into a bridge circuit and with a potential of 2 V across the bridge, they attain a temperature of ~180°C in nitrogen. There is a linear relation between the out-of-balance potential difference on the bridge and the partial pressure of hydrogen in the active cell for hydrogen pressures up to 3×10^4 Pa (0.3 atm.), corresponding to a potential difference of ~45 mV. The katharometer is calibrated empirically yielding hydrogen pressure as a function of potential. The response time for a change of hydrogen concentration is < 15 s.

Probe

The characteristics required of the probe material are:

1. Resistance to attack by molten metal.
2. Inertness to hydrogen.
3. Adequate mechanical strength and good thermal shock resistance.
4. Easy formability into complex shapes.

No completely satisfactory permanent material is available. Two materials, i.e. alumina and stainless steels, are usable within limitations. Dense pure alumina fulfils all of the requirements except for thermal shock resistance which is inadequate but greatly improved by a coating of powdered alumina, bonded to the surface by a glaze. Stainless steel fails rapidly by molten metal attack but temporary protection can be afforded by a proprietary enamel of a kind used for rocket nozzles. Neither material is wholly satisfactory and probes made from them will survive for only a limited number of immersions before replacement. For this reason, the probes are conceived as consumable components and are designed to permit easy replacement with interchangeable capillary joints to the instrument.

A probe geometry in which carrier gas bubbles are forced to break free from a bubbler tube and are collected by a hood for recirculation might seem to offer the best conditions to maintain the clean gas/metal interface needed for efficient hydrogen collection. The arrangement is impracticable because the high surface tension of the liquid metal precludes the use of any geometrical arrangement which ensures both the detachment and collection of the bubbles. As an alternative, the common transmission/reception design illustrated in Figure 5.7 is used in which a turbulent stream of the carrier gas passes along the interface between the probe and liquid metal without breaking away from it and is thereby guided into a hood. The stem of the probe is a 6 mm diameter rod with twin 1 mm diameter bores. It carries a small hood cemented in place, with 50 mm of the stem protruding below it. One bore is a vent for the outward gas stream and the other communicates with the hood space to recover the carrier gas for recirculation. A porous alumina insert in the hood prevents molten metal entering the return bore.

A general view of a simple instrument is given in Figure 5.8. The term "Telegas" refers to instruments of this general form.

5.3.3 PRACTICAL OPERATION AND CHARACTERISTICS

5.3.3.1 Operation

In taking a reading, the probe is inserted into the metal so that the hood is well immersed, i.e. about 4 cm below the surface. During entry into the metal, the carrier gas admission valve is opened, blowing a stream of nitrogen out through both the hood and bubbler tube to clear away oxide films. The valve is closed, the pump is operated until the out of balance potential is steady and the corresponding hydrogen pressure, p, is read from a calibration chart. The temperature of the metal is measured with a thermocouple close to the probe to determine the appropriate value of the solubility, S, as given for various metal compositions in Chapter 6. Inserting p and S into Eqn 5.5 yields the hydrogen content.

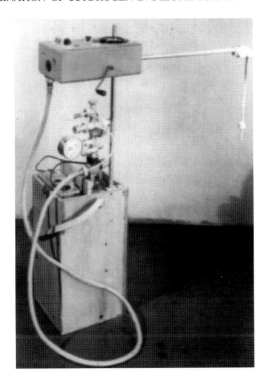

Fig. 5.8 General view of a simple telegas instrument.

The typical response curves, given in Figure 5.9, show that the response time depends on the hydrogen content, e.g. 1.5 and 3 minutes respectively for 0.11 and 0.3 cm³/100 g.

5.3.3.2 Accuracy

Results obtained with the Telegas and corresponding results by vacuum extraction for chill-cast samples taken from laboratory melts of pure aluminium are compared in Figure 5.10. There is little systematic difference between the two sets of results. Small random differences are unavoidable, because the carrier gas technique measures the hydrogen content at a position 4 cm below the surface, whereas metal for the chill sample is taken by ladle from the surface close to but not co-incident with, the Telegas probe.

When two Telegas instruments are compared with their probes side by side in the same melt their readings agree exactly.

5.3.3.3 Example of Telegas Application in Industrial Experiments

Ransley et al.[2] quote an application to acquire evidence that the hydrogen content varies within the large mass of metal in an industrial melting furnace. Twenty tonnes of 99.2% pure aluminium scrap was melted in a gas-fired furnace, the temperature was adjusted to

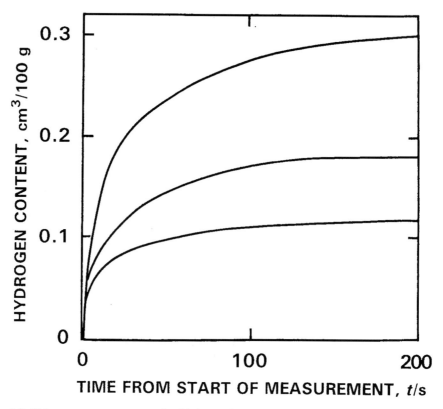

Fig. 5.9 Telegas response curves for high, moderate and low hydrogen contents in liquid pure aluminium at 700°C. (Ransley *et al.*).[2]

690°C and the metal was degassed by purging it with nitrogen applied by a carbon tube inserted through a door in the front wall. The metal was tapped from the front of the furnace into the transfer launder, where its hydrogen content was measured continuously with a Telegas instrument. The results are given in Figure 5.11 and show that the hydrogen content of the metal flowing from the furnace rose as the less-effectively degassed metal at the back of the furnace contributed to the stream. The breaks in the readings plotted in the figure correspond with resetting the instrument by purging it with carrier gas to confirm that it was functioning correctly. The solid points in the figure give the results of hydrogen contents determined by vacuum extraction on chill-cast samples taken at intervals from the metal in the launder; they agree well with the Telegas results.

5.3.3.4 Further Developments
The Telegas is generally accepted and is the subject of ongoing developments, directed towards improving its robustness and convenience for in-line industrial application as a

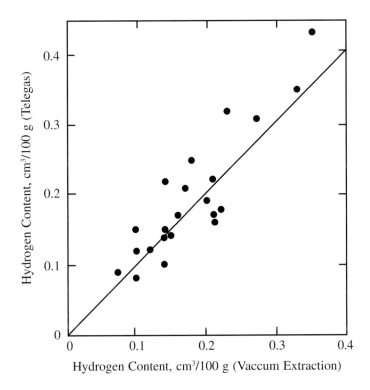

Fig. 5.10 Comparison of values of hydrogen content determined by the telegas for liquid 99.2% pure aluminium with values determined by vacuum extraction of corresponding cast samples. (Ransley *et al.*).[2]

facility for quality control.[15–18] Its derivatives include the ALSCAN instrument with a reciprocating simplified probe comprising a porous ceramic plug within which the carrier gas is recirculated[17] and the TELEGAS II incorporating a microprocessor to give readings directly in hydrogen content.[18] The basic Telegas principle on which these instruments are based remains unchanged.

5.4 RAPID CARRIER GAS METHOD

The method is an application by Degrève[3] of a rapid semi-automatic method devised by Boillot *et al.*[19] for steel. The essential features are:

1. The sample is melted by RF heating in an outgassed graphite crucible and the evolved hydrogen is collected in a flowing inert carrier gas.

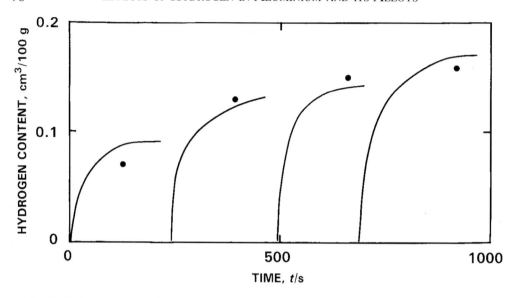

Fig. 5.11 A sequence of four telegas readings, showing a rising hydrogen content in liquid pure aluminium flowing in the launder during casting from a 20 tonne melting furnace. Solid symbols give the results of vacuum extraction determinations on corresponding cast samples for comparison. (Ransley *et al.*).[2]

2.   The carrier gas is passed once over the sample and the quantity of hydrogen evolved is determined by integrating the hydrogen content and flow of the effluent, using a simple microprocessor.

3.   The sample is prebaked by RF induction heating to ~400°C for a short period to drive off sources of spurious hydrogen on the sample surface and adjacent walls of the equipment.

4.   The sample is introduced through a screen of flowing inert gas to eliminate the need to pre-bake the furnace tube.

Various proprietary derivatives of the technique are available commercially.

5.5 APPLICATION OF REDUCED PRESSURE TO LIQUID METAL SAMPLES

These tests are simple, robust and need little service facilities. They measure the combined influences of hydrogen and bubble nuclei,[4] indicating the potential for porosity in castings poured from liquid metal of moderate quality but they are insensitive to the low hydrogen contents required for critical applications. They are of two kinds:

1. The "first bubble test".[5, 20] This test gives a crude measurement of the activity of hydrogen gas in equilibrium with hydrogen dissolved in liquid metal. The pressure over a small sample of liquid metal held at constant temperature is reduced until the first bubble is observed. Nominally, this critical pressure has the same significance as the pressure measured by the Telegas but the test is subjective and ignores the excess pressure needed for bubble nucleation.

2. The vacuum solidification[4, 21] or Straube-Pfeiffer test. In this test, a liquid metal sample is allowed to solidify either in vacuum or under a pre-determined sub-atmospheric pressure. The solid sample is sectioned and the cavitation is compared with a set of calibrated standards.

5.6 ASSESSMENT OF APPLICATIONS

Most major manufacturers maintain at least one vacuum extraction facility in central technical centres and use the rapid carrier gas method and/or the Telegas for routine measurements for quality control in production.

The vacuum extraction method is used in process development and as a control to validate or calibrate routine methods but it is inconvenient for quality control because it is slow and requires specialised laboratory support, e.g. the services of specialised technicians, provision of liquid nitrogen and on-site glass-blowing services.

Users and even suppliers of equipment for rapid carrier gas determinations are sometimes unaware of sources of error that can be introduced by complexities in the behaviour of aluminium alloys on heating in an inert atmosphere, e.g. hydrogen from spurious sources is not always completely separated from solute hydrogen evolved from samples in an arbitrary short pre-bake and there is increased potential for spurious hydrogen evolution when the metal melts. The reliability of such equipments in quality control must be carefully evaluated by comparisons with other methods, e.g. vacuum extraction. A standard deviation of 0.03 $cm^3/100$ g is claimed for metal free from volatile alloy components and impurities[22] but if chemically active volatile elements are present the method is particularly vulnerable to sources of error that do not apply to vacuum extraction:

1. Residual water vapour in imperfectly dried carrier gas.
2. Incomplete differentiation of the true hydrogen content from spurious hydrogen.
3. Accumulated contamination of the equipment by condensed volatile metals.

In principle, the Telegas is ideal for in-plant use but the probes are vulnerable in industrial environments and require periodic replacement. It is satisfactory when supported by repair and re-calibration facilities at a central service base.

Hydrogen content determinations on aluminium-zinc-magnesium alloys are difficult whatever the method used. Vacuum extraction results are subject to scatter and the carrier-gas method is unsuitable for the reasons given earlier. The Telegas is unusable because the metal penetrates through the porous ceramic into the bores of the alumina probes, blocking them, aided no doubt by wetting with an initial condensate of evaporated zinc.

5.7 REFERENCES

1. C.E. Ransley and D.E.J. Talbot, *J. Inst. Metals*, **84**, 1955-6, 445.
2. C.E. Ransley, D.E.J. Talbot and H. Barlow, *J. Inst. Metals*, **86**, 1957–58, 212.
3. F. Degrève, *Light Metals (AIME)*, **2**, 1975, 213.
4. K.J. Brondyke and P.D. Hess, *Trans AIME*, **230**, 1964, 1542.
5. Y. Dardel, *Metals Technology*, 1948, TP No. 2484 and Metals Ind., **76**, 1950, 203.
6. R.T. Fruhans, L.T. Martonik and E.T. Turkdogan, *Trans AIME*, **245**, 1969, 1501.
7. G.C. Wood, P.N. Anyalebechi and D.E.J. Talbot, *Proc. International Seminar on Refining and Alloying of Liquid Aluminium and Ferro-Alloys*, Norwegian Institute of Technology, Trondheim, 1985.
8. D.E.J. Talbot and D.A. Granger, *J. Inst. Metals*, **91**, 1962-63, 319.
9. W. Eichenauer and A. Pebler: *Z. Metallkunde*, **48**, 1957, 373.
10. W. Eichenauer, K. Hattenbach and A. Pebler, *Z. Metallkünde*, **52**, 1961, 682.
11. J. Crank, *The Mathematics of Diffusion*, Clarendon Press, Oxford, 1956.
12. C.E. Ransley and D.E.J. Talbot, *Z. Metallkünde*, **46**, 1955, 328.
13. D.E.J. Talbot and P.N. Anyalebechi, *Mater. Sci. Tech.*, **4**, 1987, 1.
14. H.A. Daynes, "*Gas Analysis by Measurement of Thermal Conductivity*", Cambridge University Press, London, 1933.
15. S. Terai, S. Sato, S. Kato, M. Imai, S. Inumaru and M. Yoshida, "Apparatus for Measuring the Hydrogen Content Dissolved in a Molten Metal", U.S. Patent No. 4 454 748, Sumitomo Light Metal Industries Ltd., 1984.
16. D.A. Granger, "Telegas for Determining Hydrogen in the Foundry Industry" *Proc. AFS/CMI Conference on Molten Metal Processing*, 1986, 417.
17. J-P. Martin, F. Tremblay and G. Dubé, *Light Metals*, 1989, 903.
18. D.A. Anderson, D.A. Granger and R.R. Avery, "The ALCOA Telegas II Instrument", *Light Metals*, 1990, 769.
19. P. Boillot, M. Hanin and C. Maeder, *L'hydrogene dans les Metaux*, International Congress, Paris, Editions Sciences et Industries, 1972, 381.
20. D.J. Niel and A.C. Burr, *Trans. AFS*, **69**, 1961, 272.
21. P.D. Hess, *J. Met.*, 1973.
22. W.D. Lamb, *Light Metals*, 1984, 1345.

6. Solubilities of Hydrogen in Aluminium and its Alloys

6.1 TECHNIQUES FOR MEASURING SOLUBILITIES

The solubilities of gases in metals are the quantities of solute in equilibrium with the gas phase at prescribed temperatures and pressures, usually quoted for 273 K and 101 kPa. Measurements are expensive because they require customised equipment and time consuming procedures. This accounts for the limited range of reliable information available. It is particularly true for hydrogen in aluminium and aluminium alloys because the quantities to be measured are small and hypersensitive to error from spurious hydrogen generated by reaction between the metal and traces of water vapour evolved within the equipment used.

The classic technique is *Sieverts' method*[1] in which a sample of the metal is exposed to the gas phase at constant temperature and pressure and the quantity of hydrogen that dissolves is determined volumetrically from the readings on a gas burette. Sieverts' method is ideally suited to determinations for liquid metals, which absorb the gas rapidly. It is less suitable for solid metals because the diffusion-controlled slow uptake of the gas prolongs the measurement, introducing unacceptable time-dependent sources of error.

A more suitable approach for solid metals is by absorption-desorption methods in which the hydrogen contents of samples are determined by vacuum extraction after equilibrating them with hydrogen. An incidental advantage of this approach is that diffusion coefficients can be calculated from gas evolution curves recorded during the vacuum extraction phase, as described in Chapter 7.

6.1.1 SIEVERTS' METHOD

A sample of the metal to be loaded with hydrogen is contained in a refractory crucible enclosed in a vacuum tight vessel, the absorption bulb, connected to a measuring system at ambient temperature. A correction is required for the volume of the gas occupying the dead space of the vessel and the connecting tubing. When the solubility of the gas is low, as it is for hydrogen in aluminium, the design of the equipment is dominated by the need to reduce the volume of the dead space to a minimum and to calibrate it accurately by blank determinations, using an insoluble gas, preferably helium. Within the constraints imposed by the design, the only material suitable for constructing the absorption bulb is fused silica. When hot, this material is permeable to hydrogen and if the gas is in contact with heated silica, as it is if external heating is used, a significant error is introduced, requiring a correction which is uncertain and cumbersome to apply.[2] The problem can be circumvented using radio frequency (RF) induction heating to maintain the sample temperature. This avoids directly heating the silica and stimulates entry of the gas into the metal by induction stirring.

6.1.1.1 Equipment

A typical form of the equipment is illustrated schematically in Figure 6.1.

The absorption bulb, A, containing the sample in its crucible is joined via small bore tubing and an isolating tap to a manifold fitted with vacuum grade taps to connect it in turn to:

1. An integral vacuum system with low-pressure gas collection facilities, not shown in Figure 6.1 but similar to that illustrated in Figure 5.1.
2. Supplies of research grade, e.g. 99.999% pure hydrogen and helium.
3. A 50 cm³ capacity gas burette, B, with a sensitivity of 0.05 cm³ filled with mercury. The volume of gas in the burette is controlled by manipulating the pressure over the mercury in a reservoir, D, using valves to air, F, and an auxiliary vacuum line, G. The pressure is measured from the difference in the mercury levels in the burette and the barometric limb, C.

The construction of the absorption bulb, illustrated in Figure 6.1b, is critically important. The sample, H, is a cylinder of the metal or alloy with a mass of about 100 g. It is machined to fit a recrystallised pure alumina crucible, J, with internal dimensions typically 55 × 40 mm diameter, allowing only sufficient clearance to accommodate the expected expansion on melting. The bulb is constructed around the crucible from concentric cylindrical outer and inner fused silica shells, K and L, each closed at one end with a flat top. With the sample in place, they are sealed at the open ends remote from the sample, M. A thermocouple, entering the system through the seal, E, is inserted into a hole, N, drilled in the sample where it is protected by a sprayed alumina coating. This construction limits the dead space of the absorption system to < 15 cm³ and allows the seal at the bottom of the bulb to be made without heating the sample.

6.1.1.2 Procedures

The absorption bulb is opened to the vacuum system, the sample is melted by RF induction heating and its temperature is controlled at the prescribed value. The first step is to precondition the sample, crucible and walls of the absorption bulb to eliminate all pre-existing sources of hydrogen by exposing them to vacuum for a considerable period, typically 2 hours. The gas evolved is monitored and when it has ceased, the absorption bulb and the high vacuum system are isolated from the manifold.

The gas burette and manifold are filled with helium at a prescribed pressure and the reading recorded. Helium is admitted from the burette to the absorption bulb by opening

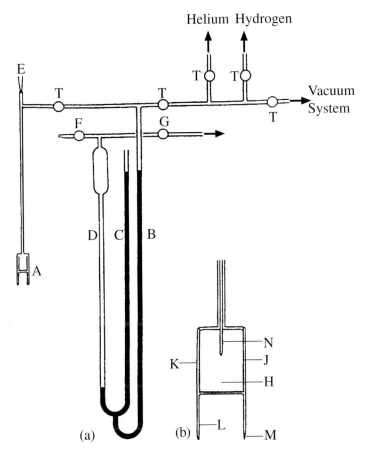

Fig. 6.1 Equipment for hydrogen solubility determinations by Sieverts' method (a) General arrangement and (b) Detail of absorption bulb.

A. Absorption Bulb	F. Valve to Air	L. Inner Silica Shell
B. Gas Burette	G. Valve to Vacuum Line	M. Seal
C. Barometric Leg	H. Sample	N. Thermocouple
D. Mercury Reservoir	J. Alumina Crucible	T. Manifold Taps
E. Thermocouple Entry	K. Outer Silica Shell	

the appropriate tap, the pressure is adjusted to its original value by manipulating the mercury reservoir and the new reading is recorded. The decrease in the volume of helium in the burette yields the equivalent dead space volume of the absorption bulb. The system is evacuated and the sequence of operations is repeated using hydrogen; the volume of hydrogen admitted exceeds the corresponding volume of helium by the volume that dissolves in the metal. Equilibrium between gaseous hydrogen and the hydrogen solute in the metal is established within about 10 minutes at all temperatures. The absorption process cannot be followed in detail because of the inertia of the mercury column. The difference between the volumes of hydrogen and of helium admitted to the bulb, corrected to standard conditions of temperature and pressure and referred to a standard mass of metal yields the solubility.

6.1.1.3 Validity of Sieverts' Method

The merit of Sievert's method is that equilibrium values are measured under equilibrium conditions and it is possible to isolate and evaluate at leisure every potential source of error by suitable supporting and blank determinations. The potential sources of error are:

1. Insensitivity due to the use of small samples or inadequacies in the design of the gas burette system used to measure the volumes of gases admitted to the absorption bulb.
2. Errors in the measurement of the hot dead space volume of the absorption bulb through thermal mismatch with hydrogen if an insoluble gas of high relative molar mass, e.g. argon, is used instead of helium.
3. Loss of hydrogen by permeation through the absorption bulb if it is directly heated.
4. Failure to reach equilibrium due to slow transport of hydrogen across the metal surface if the metal is quiescent.

The sensitivity of Sieverts' method to these sources of error when used to determine the low solubility of hydrogen in liquid aluminium was first apparent from the meticulous systematic work of Ransley and Neufeld[2] and Opie and Grant.[3]

6.1.2 ABSORPTION-DESORPTION METHODS

In these methods, samples are equilibrated with hydrogen, separated from the gas phase and submitted to vacuum extraction to determine the quantity of hydrogen dissolved. They are used in determinations for solid metal for which Sieverts' method is inappropriate by reason of the slow diffusion in the solid state. The reliability of the results depends primarily on the approach to equilibrium between the sample and the gas phase, the retention of the solute gas during quenching and effects of evaporation of volatile alloy components. These methods are unsuited to determinations for liquid metal because of difficulty in retaining dissolved hydrogen in a liquid sample for subsequent measurement of its hydrogen content.

6.1.2.1 Equipment

The form of equipment used for Sieverts' method described in Section 6.1.1 and illustrated in Figure 6.1 is well suited to the measurements. The gas burette and manifold system provide means to supply and control the gas phase during equilibration and the associated vacuum system with the facilities for low-pressure gas collection serves to measure the evolution of the solute during vacuum extraction. Clean aluminium and aluminium alloy surfaces exposed to ambient air can acquire thin coatings of hydrated oxides which yield spurious hydrogen if the metal is heated in vacuum.[2, 4] The design of the technique is therefore dominated by the need to precondition samples to dehydrate any surface films and thereafter isolate the samples from the atmosphere until a determination is completed.

6.1.2.2 Procedures

Sample Preparation

Samples of the metal must be prepared meticulously to ensure that they are free from hydrogen traps before measurements are made on them. Cast metal is unsuitable because of the risk of unsoundness. Wrought metal may contain hydrogen traps,[5] which must be eliminated before using the material for solubility determinations. This is done by prolonged annealing in high vacuum to remove all of the existing hydrogen content of the metal and heal any microscopic voids. The samples for solubility determinations are typically cylinders 10 mm diameter × 50 mm, with a mass of 8.5–11 g. They are machined from the vacuum annealed metal, removing sufficient metal from the periphery to eliminate the surface zone depleted of any volatile components of the alloy, e.g. magnesium, lithium and zinc, by evaporation during the vacuum extraction.

Equilibration

The dead space in the absorption system is not critical as it is for Sieverts' method, allowing greater freedom in construction. A silica bulb or tube containing a sample fitted with a thermocouple is assembled on the apparatus, evacuated and checked for freedom from leaks. The sample is maintained at a high temperature, preferably ≥ 873 K in vacuum for ≥ 2 hours to dehydrate any surface films on the sample and pre bake the absorption bulb by radiant heat and then allowed to cool. The criterion to proceed is that a pressure < 10^{-3} Pa is maintained indefinitely in the gas collection system when connected to the absorption bulb.

The tap to the vacuum system is closed and the gas burette, B, and manifold are filled with hydrogen. Hydrogen is admitted to the absorption bulb, A, by opening the appropriate tap. The sample is heated to and controlled at the required temperature. The hydrogen pressure in the system is adjusted and maintained at the required value by manipulating the mercury in the reservoir, D, until equilibrium with the gas phase is established. The time needed to establish equilibrium is calculated from diffusion coefficients calculated as described in Chapter 7 and applied retrospectively.

Absorption-Quench Desorption Procedure

With RF induction heating, the thermal capacity of the system does not include the reaction vessel and the sample can be quenched rapidly by switching off the RF current and applying a cold air blast to the outside of the absorption bulb, retaining almost all of the solute. Cooling curves are recorded for the sample,[6] using its internal thermocouple to assess whether any correction is needed for loss of solute during quenching. Applying

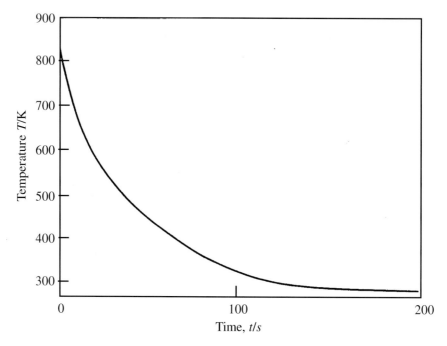

Fig. 6.2 Solubility determinations by the absorption-quench-desorption method. Typical cooling curve for 10 mm diameter aluminium alloy sample quenched by cold air applied to the outside of the absorption bulb.[6]

eqn 7.10 given in Chapter 7, finite difference calculations, guided by typical cooling curves such as that illustrated in Figure 6.2, show that the loss is < 2% of the total hydrogen content. External furnace heating is unsuitable because quenching entails moving the sample to a cool zone in the system and is less effective. The absorption bulb is evacuated and outgassed to remove hydrogen adsorbed on its walls and the hydrogen content of the sample is determined by the vacuum extraction procedure described in Chapter 5.

Isothermal Procedure
The essential feature of this version is that the three stages of the determination, i.e. equilibrating a sample with hydrogen gas, evacuating the absorption bulb to remove the gas phase and determining the hydrogen content of the sample are conducted in an unbroken sequence with the sample at constant temperature; it is accomplished by appropriate manipulation of the manifold taps. The technique relies on evacuating the absorption bulb so quickly that only a small fraction of the hydrogen dissolved in the sample is lost. The correction for lost hydrogen is determined by back extrapolation of

the hydrogen evolution curve recorded during the subsequent determination of its hydrogen content. Eichenauer and his co workers applied this method very successfully to determine the solubilities of hydrogen in solid metals,[7–9] where the desorption of hydrogen is diffusion-controlled so that the correction for lost hydrogen is small and can be determined accurately using standard diffusion theory, as explained in Chapter 7.

6.1.3 Procedures for Alloys with Active Volatile Components

Alloys containing any of the components, magnesium, zinc and lithium present additional problems due to the chemical reactivity of the metal and the evaporation of active vapours.

Alloys Containing Lithium

The problems are most acute for liquid alloys containing lithium. The only crucible materials with reasonable resistance to attack by liquid metal containing lithium even at the reduced activity in alloys are alumina, magnesia, beryllium oxide, zirconia, thoria, boron nitride, sialon and some proprietary products based on graphite. Within the constraints imposed by the design, the only practicable material for the walls of the absorption bulb is fused silica, since it can be formed by glassworking techniques into the required vacuum-tight envelopes fitting closely around the samples in situ and is resistant to high temperature and thermal shock. Lithium vapour attacks the inner surfaces of the silica walls,[10, 11, 23] yielding lithium silicide, $Li_{22}Si_5$. Where reaction is severe, as immediately above the open liquid metal surface in Sieverts' method, the reaction can penetrate through the wall thickness, producing leaks which abort determinations. The attack can be delayed to the extent that a determination can be completed by shielding the top of the bulb by a refractory cover over the liquid metal surface. Even so, it is inadvisable to use the same assembly of a sample, crucible and absorption bulb for more than one or two determinations.

A further problem is drift in alloy composition by loss of lithium from the liquid metal, which is probably the greatest single source of error. Some loss, typically 7% of the total must be accepted but it can be minimised by preconditioning the metal and heated parts of the equipment in a helium atmosphere rather than in vacuum as described in the first paragraph of Section 6.1.1.2 and limiting it to the minimum time consistent with complete removal of existing hydrogen sources.[10, 11] No error from absorption of hydrogen on lithium condensates could be detected.[10, 11]

The evaporation of lithium from some ternary and quaternary alloys can initiate an electric discharge around the sample, reducing the efficiency with which the RF power is supplied to the sample and interfering with the thermocouple reading.[11] When encountered, the discharge may be transient and clears if the RF power is reduced but for alloys with particularly high lithium activities it is persistent and the determinations are aborted. Surprisingly, the problem does not arise for binary alloys with $\leq 3\%$ lithium.[10]

As expected, lithium evaporation from samples submitted to the absorption desorption method using solid samples is a lesser problem but there is attack along lines of contact between the sample and silica that can crack the walls of the absorption system and a crucible must be provided to separate them.

Alloys Containing Magnesium or Zinc

Although magnesium and zinc vapours do not react with silica so strongly as lithium, similar precautions against attack on silica are advisable. A more serious problem is that they are so volatile that it is difficult to maintain control over alloy compositions during determinations by Sieverts' method.

6.1.4 Reliability of Determinations

Measurements of the solubility of hydrogen in aluminium alloys is time consuming, very difficult and full of traps for the unwary, requiring patience and experimental skill and observation of a high order. Transfer of expertise from the vacuum sub-fusion technique is the basis of successful application of both Sieverts' method and the absorption/ quench/extraction method to the determination of hydrogen solubilities in liquid and solid aluminium and its alloys. Experience with materials of construction for the equipment and appreciation of the natures of aluminium alloy surfaces hold the keys to reliability. For example, once a metal sample has been sealed in the equipment and both have been conditioned by heating in high vacuum, it is vitally important to exclude them from contact with any trace of water vapour or of oxygen. Thus, all gases employed must be of research grade purity, internal surfaces of the equipment must be thoroughly outgassed and liquid nitrogen cold-trapping must be applied immediately at strategic points and retained until all measurements are complete. With these precautions there is every confidence in the results; without them, the results are of dubious value.

6.2 PRESENTATION OF VALUES FOR SOLUBILITY

6.2.1 Conventional Descriptions of Solutions

The conventional quantities and symbols for solutions of hydrogen in aluminium alloys are molalities, $m_H/m_H^{\oplus}$, or mole fractions, x_H, of the monatomic solute and their dependence on pressure and temperature are expressed in eqns 4.12 and 4.13, given in Chapter 4:

$$\log\left(\frac{m_H}{m_H^{\oplus}}\right) - \tfrac{1}{2}\log\left(\frac{p}{p^{\oplus}}\right) = -\frac{\Delta H^{\oplus}}{2.303 \times 2RT} + \frac{\Delta S^{\oplus}}{2.303 \times 2R} \tag{4.12}$$

$$\log x_H - \tfrac{1}{2}\log\left(\frac{p}{p^{\oplus}}\right) = -\frac{\Delta H_{x_H}^{\oplus}}{2.303 \times 2RT} + \frac{\Delta S_{x_H}^{\oplus}}{2.303 \times 2R} \tag{4.13}$$

These descriptions are mandatory when solubilities are considered in a scientific context but they are unsuitable for use in industrial applications as explained in Section 6.2.2.

6.2.2 PRACTICAL SYSTEM BASED ON HYDROGEN CONTENT

Industrial applications require a coherent system to evaluate quantities of hydrogen contained in the metal irrespective of the states of occlusion. As a consequence, almost all measured values for solubility are published in the form of hydrogen contents, defined in Chapter 4, Section 4.4.2, i.e. the volumes of hypothetical diatomic gas equivalent to the solute in 100 g of metal measured at 273 K and 101 kPa, $V(273 \text{ K}, 101 \text{ kPa})$. Moreover, the symbol, S, is firmly established for use in equations for solubilities given in this form. This practical convention is continued in the present text, with the following stipulations:

1. S is defined as the quotient $V(273 \text{ K}, 101 \text{ kPa})/\text{cm}^3$, so that it is dimensionless and suitable for use as a subject of logarithmic operators, a nicety not always observed in publications.
2. It is printed in boldface, *viz* S, to distinguish it from the symbol, S, for entropy.

Hydrogen contents representing solubilities are usually reported as functions of temperature and pressure by fitting measured values to a derivative of eqn 4.12, obtained by substituting S for $m_H/m_H^{\ominus}$:

$$\log S - \tfrac{1}{2}\log\left(\frac{p}{p^{\ominus}}\right) = -\frac{A}{T} + B \tag{6.1}$$

selecting appropriate values for the numbers A and B.

6.2.3 RELATION OF HYDROGEN CONTENT TO MOLALITY AND MOLE FRACTION

For scientific applications, solubilities that are given in the form of hydrogen contents, S, can be transcribed to the relative molality, $m_H/m_H^{\ominus}$, or the mole fraction, x_H, of the monatomic solute, using the information:

n mole of diatomic hydrogen gas yields $2n$ mole of monatomic solute.
Molar volume of diatomic hydrogen, $V_m(273 \text{ K}, 101 \text{ kPa}) = 22400 \text{ cm}^3$.
Molar mass of aluminium, $M_{Al} = 0.027 \text{ kg}$.
Mole fraction, $x_H = n_H/(n_{Al} + n_H) \approx n_H/n_{Al}$ for a dilute solution.
Hence, selecting $m_H^{\ominus} = 1 \text{ mol kg}^{-1}$ as the standard value for molality:

$$\frac{m_H}{m_H^{\ominus}} = 2\left(\frac{S}{22400}\right) \times 10 = 8.93 \times 10^{-4} \times S \tag{6.2}$$

and:

$$x_H = \frac{n_H}{n_{Al}} = \frac{M_{Al}}{V_m \times 0.1} \times 2S = \frac{0.027}{22400 \times 0.1} \times 2S = 2.41 \times 10^{-5} \times S \qquad (6.3)$$

Substituting for S in eqn 6.1 and equating terms in $\Delta H^\ominus$ and $\Delta S^\ominus$ with the numbers A and B respectively yields eqns 4.12 or 4.13.

6.3 CRITICAL ASSESSMENTS OF PUBLISHED VALUES FOR SOLUBILITIES

6.3.1 PURE ALUMINIUM

Information on the solubilities of hydrogen in pure and liquid solid aluminium is a basic requirement for many purposes, e.g. to assess the potential of occluded hydrogen to generate porosity, to evaluate industrial degassing operations and to calibrate the Telegas instrument.

6.3.1.1 Pure Liquid Aluminium

The only values that need be considered are those reported since the refinements in equipment and procedures needed for such sensitive measurements were first appreciated by Ransley and his coworkers[2, 4] and adopted in subsequent determinations.

Four independent investigators measured the solubility as functions of temperature and pressure. Ransley and Neufeld,[2] Opie and Grant,[3] and Talbot and Anyalebechi[12] all used Sievert's method and Eichenauer *et al.*[13] used the isothermal absorption-desorption method. Values reported for 101 kPa (1 atm) are given in Table 6.1 and plotted in Figure 6.3. The three independent sets of values measured by Sievert's method all agree well and scatter evenly about a common linear isobar but the single set of values measured by the isothermal absorption-desorption method diverge systematically from the Sievert's values.

The validity of the results can be assessed against three criteria:
1. Correlation with theoretically predicted relationships;
2. Reliability of experimental methods;
3. Consistency with the measured activities of arbitrary hydrogen contents.

Theoretical Relationships

The requirements are that the results conform with the principles derived in Chapter 4 for the temperature and pressure dependence of dilute solutions of hydrogen in metals, i.e:
1. The Van't Hoff equation (eqn 4.12 or 4.13).
2. Sieverts' isotherm (eqn 4.6).

All of the determinations satisfy the Van't Hoff equation as shown by the linearity of the isobars in Figure 6.3. The pressure range for which values are available, 67 to 113 kPa, is insufficient to justify a formal Sieverts's plot but the isotherm is observed by constant

Table 6.1 Solubility of Hydrogen at 101 kPa (1 atm) in 99.99% Pure Liquid Aluminium - Measured Values

Temperature		Solubility, S (cm^3/100 g)			Isothermal Absorption & Desorption Method
		Sieverts' Method			
°C	T/K	Ransley and Neufeld[2]	Opie and Grant[3]	Talbot and Anyalebechi[12]	Eichenauer et al.[13] (Representative Values)*
670	943	0.70	-	0.62 0.69	0.49
700	973	1.10	0.90	0.92 0.93 0.94	0.61
710	983	0.97	-	-	-
725	998	0.89	-	-	0.74
730	1003	0.98 1.30	-	-	-
750	1023	1.09 1.40	-	1.25 1.31	0.88
790	1063	1.61	-	-	-
800	1073	-	1.75	1.63 1.68	1.22
820	1093	2.20	-	-	-
850	1123	1.80	-	1.82 1.89 1.97	1.65
900	1173		2.75	-	2.17
950	1223				2.78
1000	1273	-	4.15	-	3.51

*Selected from closely spaced data points given graphically in Reference 13

values of the ratio $S/(p/\text{Pa})^{1/2}$ in the example given in Table 6.2. These theoretical criteria are necessary but insufficient because they do not rule out systematic errors.

Reliability of Experimental Methods

The discrepancy between the values produced by Sieverts' method and by isothermal absorption desorption is ascribed to differences in the reliability of the two methods rather than to subjective variation between the work of different investigators, for the following reason. As explained in Section 6.1.2 the reliability of the isothermal absorption-desorption method depends on a correction for initial loss of hydrogen in the desorption stage. For solid metal, the correction is small and can be computed accurately because the loss is under diffusion control. For liquid metal the loss is greater and uncertain because it is not diffusion controlled. Eichenauer et al.[13] themselves share this reservation by confirming that the initial loss of hydrogen from their liquid samples was controlled

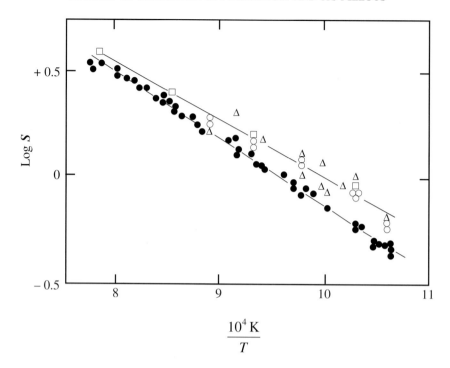

Fig. 6.3 Solubility of hydrogen, *S*, in pure liquid aluminium as a function of temperature, *T*. Van't Hoff isobars for equilibrium with the gas phase at a pressure of 101 kPa (1 atm). *S* = *V*/cm³, where *V* is the volume at 273 K and 101 kPa of diatomic hydrogen equivalent to the solute in 100 g of metal.

Δ   Measured by Sieverts' method   (Ransley and Neufeld)[2]
☐   Measured by Sieverts' method   (Opie and Grant)[3]
○   Measured by Sieverts' method   (Talbot and Anyalebechi)[12]
●   Measured by Isothermal absorption-desorption (Eichenauer *et al.*)[13]

Table 6.2 Pressure-Dependence of the Solubility of Hydrogen in Pure Liquid Aluminium at 973 K (700°C) (Talbot and Anyalebechi)[12]

p/kPa	Solubility, *S* (cm³/100 g)			$S \times (p/\mathrm{Pa})^{-1/2}$
	Replicate Values		Mean	
67	0.78	0.78	0.78	3.0×10^{-3}
80	0.81	0.82	0.82	2.9×10^{-3}
93	0.88	0.89	0.89	2.9×10^{-3}
101	0.92	0.94 0.93	0.93	2.9×10^{-3}
107	1.01	0.92	0.97	3.0×10^{-3}
113	0.99		0.99	2.9×10^{-3}

not by diffusion but by unidentified processes and they were obliged to correct for it by applying empirical "velocity constants" of an undisclosed nature.

Hydrogen Activities in Equilibrium with Arbitrary Hydrogen Contents

The solubility can be verified from the relation between arbitrary hydrogen contents and hydrogen activities manifest as pressure measured by the Telegas, explained in Chapter 5:

$$G = S \left(\frac{p}{p^{\ominus}} \right)^{\frac{1}{2}} \tag{5.5}$$

where: p is the measured pressure corresponding to an arbitrary hydrogen content, G and S is the solubility at a standard value of pressure, $p^{\ominus}$, usually 101 kPa (1 atm).

The relation can be assessed by comparing values for hydrogen contents of liquid metal measured by the Telegas with values determined by vacuum extraction on corresponding cast samples.[14-17] It is found that agreement between the measurements is good for the values of S given by Sieverts' method[2, 3, 12] but poor for the corresponding values given by the isothermal absorption-desorption method.[13]

Recommended Values

All of this evidence supports the view that the best available values for the solubility of hydrogen in pure liquid aluminium are those produced by Sieverts' method.[2, 3, 12] Scrutiny of the authors' descriptions shows that all of the work was carried out meticulously and with full appreciation of potential sources of systematic error. Since the three independent sets of points all cluster around a common isobar in Figure 6.3, it is logical to combine them by regression analysis to determine the common isobar, assigning the scatter to random errors.

This yields the following equation for the solubility at 101 kPa (1 atm):

$$\log S = - \frac{2700}{T/K} + 2.72$$

and introducing pressure as a variable, this yields the final equation:

$$\log S - \frac{1}{2} \log \left(\frac{p}{p^{\ominus}} \right) = - \frac{2700}{T/K} + 2.72 \tag{6.4}$$

6.3.1.2 Pure Solid Aluminium

The solubility of hydrogen in pure solid aluminium has been measured by the absorption desorption method on two occasions. Ransley and Neufeld[2] used the absorption-quench-desorption procedure and Eichenauer and Pebler[7] used the alternative isothermal procedure. The two sets of values are given in Table 6.3, they are self-consistent and agree well in the temperature range 520 to 620°C but diverge at lower temperatures where the solubilities are so small that they are below the sensitivity of either method.

Table 6.3 Solubility of Hydrogen at 101 kPa (1 atm) in Pure Solid Aluminium - Measured Values

Temperature		Solubility, S (cm³/100 g)	
°C	T/K	Ransley and Neufeld[2] †	Eichenauer and Pebler[7] ‡
465	738	0.010	-
470	743	0.012	0.004
475	748	0.010	-
495	768	0.013	-
500	773	-	0.006, 0.008
520	793	0.014	-
530	803	-	0.011, 0.012
535	808	0.015	-
545	818	0.017, 0.018	-
562	835	-	0.018
565	838	0.020	-
570	843	0.022	-
580	853	0.024	-
590	863	-	0.024
595	868	0.024, 0.025, 0.026	-
620	893	0.029	-

† 99.99% pure metal
‡ 99.50% pure metal

Despite the limited temperature range for which the measured values are reliable, the Van't Hoff isotherm can be fitted to them confidently with the help of the general equation for the solubility of hydrogen in an FCC metal, given in Section 4.1.2.3 as eqn 4.31:

$$\log x_H - \tfrac{1}{2}\log\left(\frac{p}{p^{\ominus}}\right) - \frac{\Delta H_x^{\ominus}}{2.303 \times 2RT} - 2.40 \tag{4.31}$$

This equation is transcribed by substituting for x_H using eqn 6.3 given in Section 6.2.3:

$$x_H = S \times 2.41 \times 10^{-5} \tag{6.3}$$

yielding: $$\log(S \times 2.41 \times 10^{-5}) - \tfrac{1}{2}\log\left(\frac{p}{p^{\ominus}}\right) = -\frac{\Delta H_x^{\ominus}}{2.303 \times 2RT} - 2.40 \tag{6.5}$$

Table 6.4 Solubility of Hydrogen at 101 kPa (1 atm) in Pure Aluminium*

Temperature		State	Solubility, S (cm³/100 g)
°C	T/K		
450	723	Solid	0.004
500	773	Solid	0.008
550	823	Solid	0.015
600	873	Solid	0.026
660	933	Solid	0.046
660	933	Liquid	0.67
700	973	Liquid	0.88
750	1023	Liquid	1.20
800	1073	Liquid	1.60
850	1123	Liquid	2.07
900	1173	Liquid	2.62

* Critically assessed values for the solid and liquid metal calculated from eqns 6.7 and 6.4 respectively.

whence:
$$\log S - \tfrac{1}{2}\log\left(\frac{p}{p^{\ominus}}\right) = -\frac{\Delta H_x^{\ominus}}{2.303 \times 2RT} + 2.22 \tag{6.6}$$

Figure 6.4 is an isobaric plot of eqn 6.6 for $p = p^{\ominus}$, fitted to the measured values extracted from Table 6.3. The best straight line through the data points is:

$$\log S = -\frac{3320}{T/K} + 2.22 \tag{6.7}$$

and reintroducing the pressure variable yields the final equation:

$$\log S - \tfrac{1}{2}\log\left(\frac{p}{p^{\ominus}}\right) = -\frac{3320}{T/K} + 2.22 \tag{6.8}$$

6.3.1.3 Comparison of Solubilities in Liquid and Solid Pure Aluminium

For convenient reference, Table 6.4 gives the critically assessed values of the solubility in the liquid and in solid metal, calculated from eqns 6.4 and 6.7 respectively, for a hydrogen pressure of 101 kPa (1 atm) at selected temperatures.

These equations yield 0.67 cm³/100 and 0.046.cm³/100 as the solubilities in the liquid and solid metal respectively at the melting point, 943 K (660°C). The solubility quotient

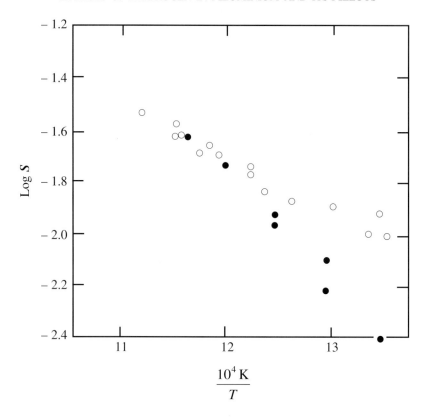

Fig. 6.4 Solubility of hydrogen, S, in pure solid aluminium as a function of temperature, T. Van't Hoff isobars for equilibrium with the gas phase at a pressure of 101 kPa (1 atm). $S = V/cm^3$, where V is the volume at 273 K and 101 kPa of diatomic hydrogen equivalent to the solute in 100 g of metal.

○   99.99% pure aluminium (Ransley and Neufeld)[2]
●   99.99% pure aluminium (Eichenauer and Pebler)[7]

at the melting point is much higher than it is for other metals, as illustrated by the examples given in Table 6.5 and it is one of the principal reasons why aluminium products are sensitive to damage by hydrogen-related artifacts, because the increase in hydrogen activity as the metal freezes generates disruptive pressures in the solid metal as explained in Chapters 9 and 10.

6.3.2 ALUMINIUM-COPPER AND ALUMINIUM-SILICON ALLOYS

Opie and Grant[3] measured the solubilities of hydrogen at 101 kPa pressure in liquid binary aluminium copper and aluminium-silicon alloys by Sieverts' method at

Table 6.5 Solubility Quotients for Hydrogen at 101 kPa (1 atm) Pressure in Some Pure Metals at their Melting Points

Metal	Melting Point (°C)	Solubility, S (cm³/100 g)		Quotient
		Liquid, S_L	Solid, S_s	S_L/S_s
Aluminium*	660	0.67	0.046	14.6
Copper[†]	1083	5.1	1.9	2.7
Iron[††]	1535	24.5	6.9	3.6
Nickel[‡]	1453	41	18	2.3
Magnesium[‡‡]	649	45	15	3.0

* From Table 6.4
[†] Röntgen and Möller,[18] Eichenauer and Pebler[7]
[††] Weinstein & Elliot,[19] Luckmayer Hasse and Schenck[20]
[‡] Schenck and Lange[21]
[‡‡] Koenman and A. J. Metcalf[22]

temperatures in the range 700–1000°C (973–1273 K) and compositions in the ranges 2–32 mass% Cu and 2–16 mass% Si. The values are probably as reliable as those for pure liquid aluminium given by the same authors quoted in Section 6.3.1.1. There is no information for solid alloys.

6.3.2.1 Liquid Aluminium Copper Alloys

The values are given in Table 6.6 and plotted as Vant' Hoff isobars in Figure 6.5, showing that copper progressively reduces the solubility. Linear regression analysis yields eqns 6.9 to 6.13.

$$\text{Al} - 2 \text{ mass%Cu} \quad \log S - \tfrac{1}{2}\log\left(\frac{p}{p^{\ominus}}\right) = -\frac{2950}{T/K} + 2.90 \tag{6.9}$$

$$\text{Al} - 4 \text{ mass%Cu} \quad \log S - \tfrac{1}{2}\log\left(\frac{p}{p^{\ominus}}\right) = -\frac{3050}{T/K} + 2.94 \tag{6.10}$$

$$\text{Al} - 8 \text{ mass%Cu} \quad \log S - \tfrac{1}{2}\log\left(\frac{p}{p^{\ominus}}\right) = -\frac{3150}{T/K} + 2.94 \tag{6.11}$$

$$\text{Al} - 16 \text{ mass%Cu} \quad \log S - \tfrac{1}{2}\log\left(\frac{p}{p^{\ominus}}\right) = -\frac{3150}{T/K} + 2.83 \tag{6.12}$$

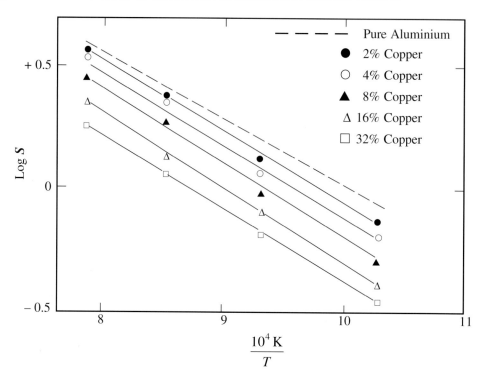

Fig. 6.5 Solubility of hydrogen, *S*, in liquid aluminium-copper alloys as a function of temperature, *T*. Van't Hoff isobars for equilibrium with the gas phase at a pressure of 101 kPa (1 atm). (Opie and Grant).[3]

$S = V/cm^3$, where *V* is the volume at 273 K and 101 kPa of diatomic hydrogen equivalent to the solute in 100 g of metal.

Table 6.6 Solubility of Hydrogen at 101 kPa (1 atm) in Liquid Binary Aluminium Copper Alloys*

Temperature		Al – 2 mass% Cu	Al – 4 mass% Cu	Al – 8 mass% Cu	Al – 16 mass% Cu	Al – 32 mass% Cu
°C	T/K	Solubility, S (cm³/100 g)	Solubility, S (cm³/100 g)	Solubility, S (cm³/100 g)	Solubility, S (cm³/100 g)	Solubility, S (cm³/100 g)
700	973	0.75	0.65	0.50	0.40	0.35
800	1073	1.35	1.15	0.95	0.80	0.65
900	1173	2.45	2.25	1.85	1.35	1.15
1000	1273	3.80	3.45	2.95	2.30	1.80

*Opie and Grant[3]

Table 6.7 Solubility of Hydrogen at 101 kPa (1 atm) in Liquid Binary Aluminium Silicon Alloys*

Temperature		Al – 2 mass% Si	Al – 4 mass% Si	Al – 8 mass% Si	Al – 16 mass% Si
°C	T/K	Solubility, S (cm³/100 g)	Solubility, S (cm³/100 g)	Solubility, S (cm³/100 g)	Solubility, S (cm³/100 g)
700	973	0.75	0.70	0.60	0.50
800	1073	1.50	1.35	1.25	1.23
900	1173	2.50	2.35	2.25	2.15
1000	1273	3.90	3.75	3.60	2.37

*Opie and Grant[3]

$$Al - 32\ mass\%Cu \quad \log S - \tfrac{1}{2}\log\left(\frac{p}{p^{\ominus}}\right) = -\frac{3150}{T/K} + 2.57 \tag{6.13}$$

6.3.2.2 Liquid Aluminium-Silicon Alloys

The values are given in Table 6.7[3] and plotted as Vant' Hoff isobars in Figure 6.6, showing that silicon also progressively reduces the solubility. Linear regression analysis yields eqns 6.14 to 6.17.

$$Al - 2\ mass\%Si \quad \log S - \tfrac{1}{2}\log\left(\frac{p}{p^{\ominus}}\right) = -\frac{2800}{T/K} + 2.79 \tag{6.14}$$

$$Al - 4\ mass\%Si \quad \log S - \tfrac{1}{2}\log\left(\frac{p}{p^{\ominus}}\right) = -\frac{2950}{T/K} + 2.91 \tag{6.15}$$

$$Al - 8\ mass\%Si \quad \log S - \tfrac{1}{2}\log\left(\frac{p}{p^{\ominus}}\right) = -\frac{3050}{T/K} + 2.95 \tag{6.16}$$

$$Al - 16\ mass\%Si \quad \log S - \tfrac{1}{2}\log\left(\frac{p}{p^{\ominus}}\right) = -\frac{3150}{T/K} + 3.00 \tag{6.17}$$

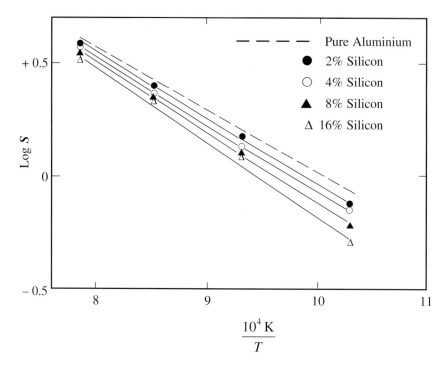

Fig. 6.6 Solubility of hydrogen, *S*, in liquid aluminium–silicon alloys as a function of temperature, *T*. Van't Hoff isobars for equilibrium with the gas phase at a pressure of 101 kPa (1 atm). (Opie and Grant).[3]

$S = V/cm^3$, where *V* is the volume at 273 K and 101 kPa of diatomic hydrogen equivalent to the solute in 100 g of metal.

6.3.3 ALUMINIUM–LITHIUM ALLOYS

Solubility measurements were initiated to obtain information needed to avert defects generated by hydrogen in products for aerospace applications but they led to the discovery of unexpected new phenomena. Solutions of hydrogen in aluminium-lithium alloys in the *liquid* state exhibit the classic features of simple endothermic solution described in Chapter 4 but corresponding solutions in the *solid* state are much more concentrated and complex than equivalent solutions in pure aluminium and other alloys and require elaboration of the theoretical infrastructure to explain their characteristics.

Solubilities of hydrogen have been determined for the experimental binary alloys, Al – 1 mass% Li, Al – 2 mass% Li and Al – 3 mass% Li and the multicomponent commercial alloys, AA 2090, AA 2091, AA 8090 and AA 8091, with the compositions given in Table 6.8. The values reported for the liquid and solid alloys are collected in Sections 6.3.3.1 and 6.3.3.3 respectively.

Table 6.8 Aluminium–Lithium Alloys for which Hydrogen Solubilities are Known

Specification	%Li	%Cu	%Mg	%Zr	%Si	%Fe	%Na	% Others
Al – 1 mass% Li	1.02	0.000	0.000	0.001	0.01	0.03	0.0001	Undetected
Al – 2 mass% Li	2.01	0.002	0.001	0.001	0.06	0.07	0.0007	Undetected
Al – 3 mass% Li	3.02	0.001	0.001	0.001	0.01	0.05	0.0001	Undetected
AA 2090 Alloy	2.14	2.60	0.01	0.11	0.04	0.06	0.003	>0.03
AA 2091 Alloy	2.10	2.10	1.53	0.11	0.03	0.07	0.0007	>0.03
AA 8090 Alloy	2.33	1.06	0.64	0.12	0.04	0.05	4 ppm	>0.03
AA 8091 Alloy	2.44	1.74	0.71	0.12	0.04	0.06	> 1 ppm	>0.03

6.3.3.1 Liquid Aluminium–Lithium Alloys

Table 6.9 gives values of the solubilities in liquid binary alloys and AA 2090 alloy as functions of temperature in the range 670 – 800°C (943 – 1073 K) at a constant hydrogen pressure, 101 kPa.[10, 11] They are plotted in the format of Vant' Hoff isobars in Figure 6.7.

Table 6.10 gives values as functions of hydrogen pressure[10] in the range 53 to 107 kPa (0.5 to 1.1 atm) at 700°C (973 K). They yield the Sieverts' isotherms in Figure 6.8.

Linear regression analysis of these values yields eqns 6.18 to 6.21.

$$\text{Al} - 1\,\text{mass\%Li}: \quad \log S - \tfrac{1}{2}\log\left(\frac{p}{p^{\ominus}}\right) = -\frac{2133}{T/K} + 2.568 \tag{6.18}$$

$$\text{Al} - 2\,\text{mass\%Li}: \quad \log S - \tfrac{1}{2}\log\left(\frac{p}{p^{\ominus}}\right) = -\frac{2797}{T/K} + 3.229 \tag{6.19}$$

$$\text{Al} - 3\,\text{mass\%Li}: \quad \log S - \tfrac{1}{2}\log\left(\frac{p}{p^{\ominus}}\right) = -\frac{2889}{T/K} + 3.508 \tag{6.20}$$

$$\text{AA\,2090\,Alloy}: \quad \log S - \tfrac{1}{2}\log\left(\frac{p}{p^{\ominus}}\right) = -\frac{2243}{T/K} + 3.019 \tag{6.21}$$

Table 6.9 Temperature-Dependence of Solubility of Hydrogen at 101 kPa (1 atm) in Liquid Aluminium–Lithium Alloys

Temperature		Al – 1 mass% Li		Al – 2 mass% Li		Al – 3 mass% Li		AA2090 Alloy	
		Solubility, S/(cm³/100 g)		Solubility, S/(cm³/100 g)		Solubility, S/(cm³/100 g)		Solubility, S/(cm³/100 g)	
°C	T/K	Replicate Values*	Mean	Replicate Values*	Mean	Replicate Values*	Mean	Replicate Values†	Mean
670	943	1.90 2.13 2.17	2.07	-	-	2.65 2.81	2.73	-	
675	948	-	-	-	-	-	-	4.37 4.48 4.49	4.45
700	973	2.46 2.52 2.62	2.53	{2.75 2.78 2.78 / 2.79 2.80 2.92}	2.80	3.47 3.53 3.66	3.55	4.90 5.39 5.44	5.24
725	998	-	-	{3.24 3.47 3.49 / 3.52}	3.43	3.81	3.81	6.03 6.09	6.06
750	1023	3.26 3.40	3.33	{3.90 4.02 4.07 / 4.09}	4.02	4.79	4.79	6.53 6.66	6.60
775	1048	3.52 3.55 3.65	3.57	{4.13 4.46 4.66 / 4.68}	4.48	-	-	7.59 7.67	7.63
800	1073	3.79 3.91	3.85	{5.17 5.30 5.36 / 5.43 5.50}	5.35	6.47 6.53	6.50	-	-

*Anyalebechi et al.[10]   †Sargent[11]

Table 6.10 Pressure-Dependence of Solubility of Hydrogen in Liquid Aluminium–Lithium Alloys at 973 K (700°C)

Pressure p/kPa	Al – 1 mass% Li Solubility, S/(cm³/100 g)		Al – 2 mass% Li Solubility, S/(cm³/100 g)		Al – 3 mass% Li Solubility, S/(cm³/100 g)		AA 2090 Alloy Solubility, S/(cm³/100 g)	
	Replicate Values*	Mean	Replicate Values*	Mean	Replicate Values*	Mean	Replicate Values[†]	Mean
53	1.90 1.89 1.94	1.91	2.17 2.32 2.10 2.05	2.16	-	-	3.70 3.60	3.65
67	2.13 2.29	2.21	2.24	2.24	2.77 2.91 2.87	2.85	4.25 4.40	4.33
76	-	-	2.55	2.55	-	-	4.29	4.29
80	2.45	2.45	2.46	2.46	3.14 3.31	3.23	-	-
87	-	-	2.69 2.53 2.50 2.40	2.53	-	-	5.01 4.90	4.96
93	2.52 2.42	2.47	2.66	2.66	3.52	3.52	-	-
101	2.52 2.62 2.46	2.53	2.78 2.79 2.78 2.80 2.75	2.78	3.47 3.66 3.53	3.55	4.90 5.39 5.44	5.24
107	2.73 2.78	2.76	2.86	2.86	3.87 3.90	3.89	5.42	5.42

*Anyalebechi et al.[10]   †Sargent[11]

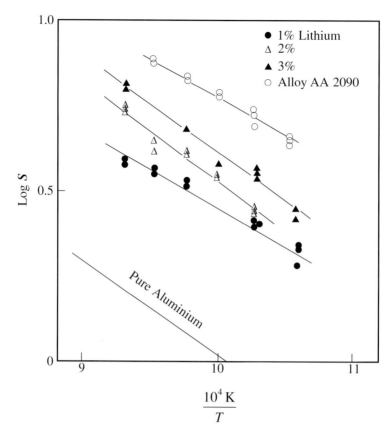

Fig. 6.7 Solubility of hydrogen, S, in liquid aluminium-lithium alloys as functions of temperature, T. Van't Hoff isobars for equilibrium with the gas phase at a pressure of 101 kPa (1 atm). $S = V/cm^3$ where V is the volume at 273 K and 101 kPa of diatomic hydrogen equivalent to the solute in 100 g of metal. (1, 2 and 3% Li binary alloys Anyalebechi *et al.*,[10] AA 2090 Sargent.[11] Corresponding plot for pure aluminium derived from eqn 6.4).

Validity of Measured Values for the Liquid Alloys

The measured values satisfy the following criteria for reliability:

Reliability of experimental methods - All of the values were measured by the proven Sieverts' method. Detailed descriptions by the authors, Anyalebechi *et al.*,[10] Sargent[11] and McCracken,[23] show that the equipment and procedures were designed to minimise expected sources of error, in accordance with best practices critically reviewed in Chapter 6, Section 6.1.1.

    Reproducibility of results - There are insufficient values for statistical analysis but replicates given in Tables 6.9 and 6.10 agree well in spite of the difficulties due to the volatility of lithium and the corrosive nature of the vapour, considered in Section 6.1.3.

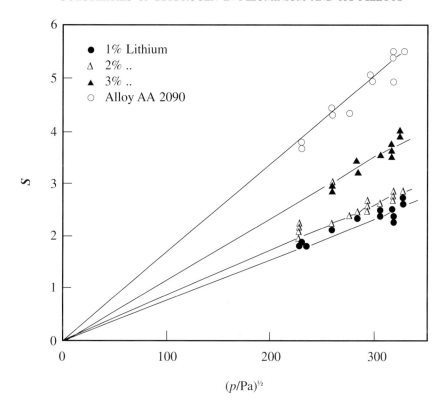

Fig. 6.8 Solubility of hydrogen, S, in liquid aluminium–lithium alloys as functions of pressure, p. Sieverts' isotherms for equilibrium with the gas phase at 973 K (700°C). $S = V/cm^3$, where V is the volume at 273 K and 101 kPa of diatomic hydrogen equivalent to the solute in 100 g of metal. (1, 2 and 3% Li binary alloys Anyalebechi *et al.*,[10] AA 2090 Sargent).[11]

Results cannot be validated by measuring activities with the Telegas instrument, because alloys containing lithium block the Telegas probes.

Correlation with theoretically predicted relationships – The linearity of the plots of the measured values of solubility as functions of temperature in Figure 6.7 and of pressure in Figure 6.8 show that they satisfy both the Vant' Hoff isobar and Sievert's isotherm.

6.3.3.2 Nature of Hydrogen Solutions in Liquid Aluminium Alloys Containing Lithium

The variations of the solubilities with temperature and pressure comply with the features expected for simple endothermic solution of a diatomic gas, described in Chapter 4. The

Van't Hoff isobars, given in Figure 6.7, are linear with negative slopes and the Sieverts' isotherms, given in Figure 6.8, are also linear and extrapolate to the origin.

Thermodynamic Properties

The solubility of hydrogen in liquid aluminium-lithium alloys is higher than that in pure liquid aluminium and rises with increasing lithium content implying that the standard Gibbs free energy change for dissolution of hydrogen, $\Delta G^{\ominus}$, becomes more and more negative. By definition, standard free energy is the sum of corresponding enthalpy and entropy terms:

$$\Delta G^{\ominus} = \Delta H^{\ominus} - T\Delta S^{\ominus}$$

The enthalpy change, $\Delta H^{\ominus}$, is determined by the binding energy for the monatomic solute hydrogen within the liquid metal and the entropy is determined by the configuration of the solute atoms within the liquid structure and the distribution of energy among them.

On first consideration, it might seem logical to attribute the influence of lithium in raising the solubility primarily to an increase in the binding energy, in view of the strong affinity of lithium for hydrogen at lower temperatures. This proposition can be assessed using eqns 6.18 to 6.21, that contain information on the separate contributions of the enthalpies and entropies of formation for the solutions they represent. They are particular examples of eqn 6.1, a derivative of the Vant' Hoff isotherm given in Section 6.2.2.

$$\log S - \tfrac{1}{2}\log\left(\frac{p}{p^{\ominus}}\right) = -\frac{A}{T/K} + B \tag{6.22}$$

which contains the following information:
1.   The slope, A, yields the standard enthalpy of solution, $\Delta H^{\ominus}$, by matching terms in eqns 6.1 and 6.22:

$$A = \frac{\Delta H^{\ominus}}{2R}$$

where: $\Delta H^{\ominus}$(diatomic H_2 gas) = 2.303 × 2 × R × A   (6.23)

Binding enthalpies for the monatomic solute hydrogen are then obtained by taking half of the result of subtracting the enthalpy of dissociation of diatomic hydrogen which is virtually constant at 419500 J mol^{-1} within the temperature range of interest:

$$\Delta H^{\ominus} \text{ (monatomic H solute)} = \tfrac{1}{2}\{\Delta H^{\ominus} \text{ (diatomic } H_2 \text{ gas)} - 419500\} \tag{6.24}$$

Enthalpies calculated in this way from eqns 6.18 to 6.21 are given in Table 6.11. They show that addition of lithium to liquid aluminium does not influence the solute binding enthalpy very much.
2.   The constant term, B, is directly proportional to the entropy of solution. It does not yield *absolute* values, because the diatomic gas and monatomic solute are referred to different standard states, but values of B for the various alloys yield *relative* values.

Table 6.11 Standard Enthalpies of Solution, $\Delta H^{\ominus}$, for Hydrogen in Liquid Binary Aluminium–Lithium Alloys and AA 2090 Alloy Compared with Values for Pure Aluminium and an Al–2.5 mass% Copper Alloy

Alloy	Diatomic Hydrogen (ΔH/kJ mol^{-1})	Monatomic Hydrogen (ΔH/kJ mol^{-1})
Pure Al	103.0[§]	−158.3
Al–1 mass% Li	80.9[*]	−169.3
Al–2 mass% Li	107.1[*]	−156.2
Al–3 mass% Li	110.0[*]	−154.5
Al–2.5 mass% Cu	113.9[‡]	−152.8
AA 2090	85.9[†]	−166.8

[*]Anyalebechi *et al.*[10] [†]Sargent[11] [‡]Opie and Grant[3] [§]Talbot and Anyalebechi[12]

The enthalpies and entropies of solution for hydrogen in the various alloys containing lithium can be assessed qualitatively by comparing the isobars given in Figure 6.7, with the appended isobar for pure liquid aluminium. As the lithium content of the binary alloys increases, the isobars are progressively displaced upwards but without significant change in slope. Thus the constant, B, in eqn 6.22 increases while the slope, A remains constant, showing that lithium raises the solubility by increasing the entropy of solution rather than by increasing the solute binding energy. For the reasons given in Section 4.1.2.3, Chapter 4, added components can raise the entropy of solution for hydrogen only by generating additional favourable sites for hydrogen atoms within the liquid metal structure. Liquid aluminium–lithium alloys exhibit negative deviation from Raoult's law[25] so that it may be lithium-rich clusters that afford the additional sites.

The solubility of hydrogen in the ternary aluminium–lithium–copper alloy, AA 2090, represented by its isobar in Figure 6.7, is much higher than that in its binary counterpart with virtually the same lithium content, i.e. 2 mass%. The isobar is displaced upwards without change in slope, showing that the solubility is raised by increasing the entropy of solution. This contrasts with the effect of copper in binary aluminium copper alloys[3] in which it depresses the solubility of hydrogen as illustrated in Figure 6.5. This implies that lithium and copper interact synergistically to produce additional sites for hydrogen when both are present, perhaps in clusters foreshadowing the tendency to form the ternary compounds, T_1 (Al_2CuLi) and T_2 (Al_6CuLi_3) that are stable in the solid alloy.

6.3.3.3 Solid Aluminium–Lithium Alloys

Tables 6.12 to 6.14 give solubilities of hydrogen determined as functions of temperature and pressure[6, 11, 23] by the absorption-quench-desorption method.

Table 6.12 Temperature-Dependence of Solubility of Hydrogen at 101 kPa (1 atm) in Solid Binary Aluminium–Lithium Alloys

Temperature		Al – 1 Mass% Li		Al – 2 Mass% Li		Al – 3 Mass% Li	
		Solubility, S/(cm^3/100 g)		Solubility, S/(cm^3/100 g)		Solubility, S/(cm^3/100 g)	
°C	T/K	Replicate Values	Mean	Replicate Values	Mean	Replicate Values	Mean
200	473	0.67* 0.67*	0.67	1.05* 1.01†	1.03	1.22* 1.26*	1.24
225	498	0.72*	0.60	1.24* 1.22* 1.12† 1.05†	1.16	2.68* 2.73*	2.71
250	523	0.78* 0.74*	0.76	1.12† 1.20†	1.19	1.36*	1.36
275	548	0.83*	0.83	1.25*	1.25		
300	573	0.89* 0.89*	0.89	1.29* 1.32* 1.29†	1.30	1.58* 1.60* 1.65*	1.61
325	598	0.97*	0.97	1.38*	1.38		
350	623	1.03* 0.97*	1.00	1.43* 1.38† 1.34†	1.38	1.64* 1.67*	1.66
375	648	1.05*	1.05	1.45*	1.45		
400	673	0.98* 0.95*	0.97	1.46†	1.46	2.31* 2.30*	2.31
425	698	0.58* 0.64*	0.61	1.72* 1.59*	1.66		
450	723	0.59* 0.61*	0.60	1.60* 1.64* 1.56* 1.66† 1.66† 1.70†	1.64	2.68* 2.73*	2.71
475	748	-	-	0.75* 0.69†	0.72	-	-
500	773	0.69* 0.68*	0.69	0.85* 0.82* 0.80†	0.82	1.82*	1.82
525	798	0.71*	0.71	0.81* 0.81† 0.83†	0.81	1.26* 1.46* 1.36*	1.36
550	823	0.79*	0.79	0.81 0.93* 0.90* 0.88† 0.94†	0.89	1.40* 1.39*	1.40
575	848	0.81*	0.81	0.86* 0.85* 0.95†	0.89	1.32* 1.64* 1.72* 1.56*	
600	873	0.83* 0.84* 0.85*	0.84	1.04* 0.99* 1.00* 1.05†	1.02	1.62* 1.63*	1.63

*Anyalebechi et al.[6]   †Sargent[11]

Table 6.13 Temperature-Dependence of Solubility of Hydrogen at 101 kPa (1 atm) in Solid Multicomponent Alloys Containing Lithium

| Temperature | | AA 2090 Alloy | | AA 2091 Alloy | | AA 8090 Alloy | | AA 8091 Alloy | |
| °C | T/K | Solubility, S/(cm³/100 g) | | Solubility, S/(cm³/100 g) | | Solubility, S/(cm³/100 g) | | Solubility, S/(cm³/100 g) | |
		Replicate Values	Mean	Replicate Values	Mean	Replicate Values	Mean	Replicate Values	Mean
200	473	1.22† 1.26†	1.24	1.29† 1.34†	1.32	-	-	-	-
225	498	1.38†	1.38	-	-	0.58‡ 0.61‡	0.60	-	-
250	523	1.37‡ 1.50‡ 1.62‡ 1.67†	1.54	1.55‡	1.55	0.96‡ 0.85‡ 0.75‡	0.85	0.80‡	0.80
300	573	1.62† 1.54†	1.58	1.56† 1.76† 1.62†	1.65	1.58‡	1.58	2.23‡	2.23
325	598	1.63‡ 1.71‡ 1.93‡	1.76		-	1.78‡ 1.55‡ 2.02‡	1.78	-	-
350	623	2.05†	2.05	1.95† 2.00†	1.98	2.50‡ 2.41‡	2.46	3.04‡	3.04
375	648	1.75†	1.75		-	-	-	-	-
400	673	2.14† 2.16† 1.86† 2.57‡	2.15	2.30† 2.06†	2.12	3.77‡ 3.39‡	3.58	4.59‡	4.59
435	708	2.14†	2.14		-		-	-	-
450	723	2.58† 2.40†	2.49	2.47† 2.47†	2.47	5.11‡ 4.42‡	4.77	5.42‡	5.42
475	748	2.59†	2.59		-	4.95‡	4.95	6.54‡	6.54
500	773	1.96† 2.11†	2.04	1.84† 2.29† 1.77†	1.97	2.90‡ 2.61‡ 1.56‡	2.36	5.13‡ 4.00‡	4.57
510	783	-	-	-	-	0.31‡ 0.56‡	0.44	2.18‡	2.18
515	788	-	-	-	-	1.18‡	1.18	-	-
520	793	-	-	-	-	-	-	1.56‡	1.56
525	798	1.16† 1.17† 1.12† 1.20†	1.16	-	-		-	-	-
530	803	-	-	-	-	0.53‡	0.53	-	-
550	823	1.39†	-	1.39	-	-	-	2.00‡ 1.96‡ 2.07‡	2.02
570	843	-	-	-	-	0.91‡	0.91	-	-
575	848	1.35† 1.42†	1.39	-	-	-	-	-	-
600	873	1.29†	1.29	-	-	-	-	-	-
620	893	1.48† 1.50†	1.49	-	-	-	-	-	-

†Sargent[11] ‡McCracken[23]

Table 6.14 Pressure-Dependence of Solubility of Hydrogen in Solid Aluminium–Lithium Alloys

Temperature °C	T/K	Pressure p/kPa	Al-1 Mass% Li Solubility*, S/(cm³/100 g)	Al-2 Mass% Li Solubility*, S/(cm³/100 g)	Al-3 Mass% Li Solubility*, S/(cm³/100 g)	AA 2090 Alloy Solubility†, S/(cm³/100 g)	AA 2091 Alloy Solubility†, S/(cm³/100 g)
300	573	26.7	0.43	0.69	0.97	0.90	0.86
300	573	53.3	0.64	1.03	1.57	1.18	1.18
300	573	66.7	–	–	–	1.40	1.52
300	573	80.0	0.67	1.18	1.63	1.54	1.45
300	573	93.3	0.72	1.28	1.82	1.60	–
300	573	101.3	0.89	1.31	1.59	1.58	1.65
300	573	106.7	0.86	1.38	2.18	1.63	1.83
300	573	109.3	0.91	–	2.36	–	–
300	573	113.3	–	1.40	–	1.69	–
600	873	26.7	–	0.52	0.84	–	–
600	873	53.3	0.58	0.73	1.23	–	–
600	873	66.7	–	0.82	1.38	–	–
600	873	80.0	0.72	0.89	1.49	–	–
600	873	93.3	0.80	0.97	1.71	–	–
600	873	101.3	0.84	1.02	1.63	–	–
600	873	106.7	0.85	1.00	1.71	–	–
600	873	109.3	–	–	1.88	–	–
600	873	113.3	0.91	1.10	–	–	–

*Anyalebechi et al.[10]   †Sargent[11]

Analysis of data points in Tables 6.12 to 6.14 yields Equations 6.25 to 6.37.

Al - 1% Li

$$473 < T/K < 680: \quad \log S - \tfrac{1}{2}\log\left(\frac{p}{p^{\ominus}}\right) = \frac{-358}{T/K} + 0.576 \tag{6.25}$$

$$680 < T/K < 873: \quad \log S - \tfrac{1}{2}\log\left(\frac{p}{p^{\ominus}}\right) = \frac{-604}{T/K} + 0.620 \tag{6.26}$$

Al - 2% Li

$$473 < T/K < 740: \quad \log S - \tfrac{1}{2}\log\left(\frac{p}{p^{\ominus}}\right) = \frac{-273}{T/K} + 0.597 \tag{6.27}$$

$$740 < T/K < 873: \quad \log S - \tfrac{1}{2}\log\left(\frac{p}{p^{\ominus}}\right) = \frac{-676}{T/K} + 0.767 \tag{6.28}$$

Al - 3% Li

$$523 < T/K < 770: \quad \log S - \tfrac{1}{2}\log\left(\frac{p}{p^{\ominus}}\right) = \frac{-615}{T/K} + 1.27 \tag{6.29}$$

$$770 < T/K < 873: \quad \log S - \tfrac{1}{2}\log\left(\frac{p}{p^{\ominus}}\right) = \frac{-830}{T/K} + 1.17 \tag{6.30}$$

AA 2090 alloy

$$473 < T/K < 773: \quad \log S - \tfrac{1}{2}\log\left(\frac{p}{p^{\ominus}}\right) = \frac{-376}{T/K} + 0.889 \tag{6.31}$$

$$773 < T/K < 893: \quad \log S - \tfrac{1}{2}\log\left(\frac{p}{p^{\ominus}}\right) = \frac{-714}{T/K} + 0.971 \tag{6.32}$$

AA 2091 alloy

$$473 < T/K < 773: \quad \log S - \tfrac{1}{2}\log\left(\frac{p}{p^{\ominus}}\right) = \frac{-366}{T/K} + 0.880 \tag{6.33}$$

AA 8090 alloy

$$498 < T/K < 773: \quad \log S - \tfrac{1}{2}\log\left(\frac{p}{p^{\ominus}}\right) = \frac{-1420}{T/K} + 2.618 \tag{6.34}$$

$$773 < T/K < 843: \quad \log S - \tfrac{1}{2}\log\left(\frac{p}{p^{\ominus}}\right) = \frac{-3630}{T/K} + 4.250 \tag{6.35}$$

AA 8091 alloy

$$523 < T/K < 773: \quad \log S - \tfrac{1}{2}\log\left(\frac{p}{p^{\ominus}}\right) = \frac{-1470}{T/K} + 2.779 \tag{6.36}$$

$$773 < T/K < 843: \quad \log S - \tfrac{1}{2}\log\left(\frac{p}{p^{\ominus}}\right) = \frac{-2390}{T/K} + 3.182 \tag{6.37}$$

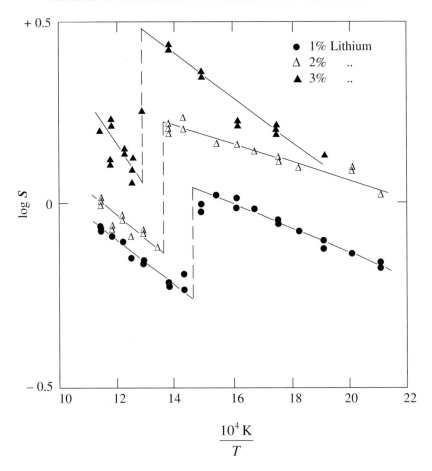

Fig. 6.9 Solubilities of hydrogen, S, in solid binary aluminium–lithium alloys as functions of temperature, T. Van't Hoff isobars for equilibrium with the gas phase at a pressure of 101 kPa (1 atm). (Anyalebechi *et al.*).[6]
$S = V/cm^3$, where V is the volume at 273 K and 101 kPa of diatomic hydrogen equivalent to the solute in 100 g of metal.

The solubility in every alloy is described by a pair of equations, applicable above and below a discontinuity at an alloy-specific critical temperature. This duplex form is unusual and is illustrated in Figures 6.9 and 6.10, where the equations are plotted as Vant' Hoff isobars for $p = p^\oplus = 101$ kPa, together with the corresponding data points from Tables 6.12 to 6.14. Table 6.15 gives a list of critical temperatures for the alloys. Table 6.16 gives standard enthalpies of solution for the monatomic gas derived from equations 6.25 to 6.37. They are more negative than the value for pure aluminium, showing that the lithium increases the binding enthalpy for the solute.

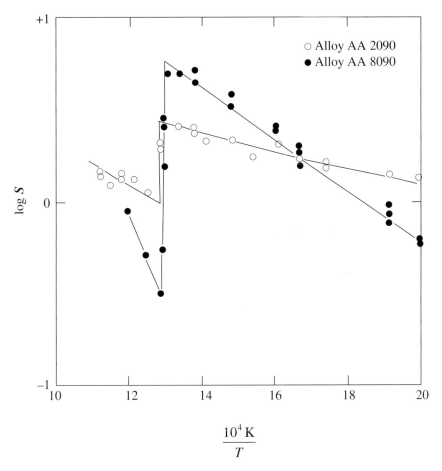

Fig. 6.10 Solubilities of hydrogen, S, in solid multicomponent aluminium–lithium alloys as functions of temperature, T. Van't Hoff isobars for equilibrium with the gas phase at a pressure of 101 kPa (1 atm) (AA 2090 Sargent[11] AA 8090 McCracken).[23]
$S = V/\text{cm}^3$, where V is the volume at 273 K and 101 kPa of diatomic hydrogen equivalent to the solute in 100 g of metal.

Tables 6.12 and 6.13 show that the solubilities of hydrogen in solid single phase aluminium alloys containng 1 to 3 mass% lithium are more than an order of magnitude greater than the corresponding solubilities in solid pure aluminium given earlier in Table 6.4. The solubilities increase with rising lithium content and are highest in commercial alloys that contain copper and magnesium as well as lithium. Complementary investigations described later in this chapter yield information that can explain the essential features of the enhanced solubility in terms of favourable sites for hydrogen associated with lithium-rich clusters in the α-phase.

Table 6.15 Critical Temperature in Vant' Hoff Isobars for Various Aluminium – Lithium Alloys

Alloy	Critical Temperature	
	°C	*T*/K
Al–1 mass% Li	407	680
Al–2 mass% Li	467	740
Al–3 mass% Li	497	770
Alloy AA 2090	500	773
Alloy AA 2091	Not Detected	
Alloy AA 8090	500	773
Alloy AA 8091	500	773

Table 6.16 Standard Enthalpies of Solution, $\Delta H^{\ominus}$ for Hydrogen in Solid Aluminium–Lithium Alloys

Alloy	Temperature Range *T*/K	$\Delta H^{\ominus}$/(kJ mol^{-1})	
		Diatomic Hydrogen	Monatomic Hydrogen
Pure Al[*]	623–933	79.4	−170.0
Al–1 mass% Li[†]	473–680	13.7	−202.9
	680–873	23.1	−198.1
Al–2 mass% Li[†]	473–740	10.4	−204.6
	740–873	25.9	−196.8
Al–3 mass% Li[†]	523–770	23.6	−198.0
	770–873	31.8	−193.9
Alloy AA 2090[‡]	473–773	14.4	−202.4
	773–893	27.3	−196.0
Alloy AA 2091[‡]	473–773	14.0	−202.7

[*]Eichenhauer and Pebler[7]   [†]Anyalebechi *et al.*[6]   [‡]Sargent[11]

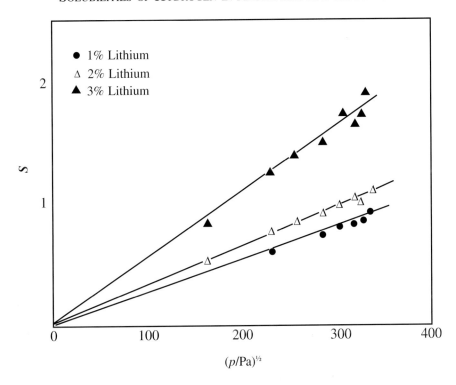

Fig. 6.11 Solubilities of hydrogen, *S*, in solid binary aluminium-lithium alloys as functions of pressure, *p*. Sieverts' isotherms for equilibrium with the gas phase at 873 K (600°C). (Anyalebechi *et al.*).[6]
S = *V*/cm³, where *V* is the volume at 273 K and 101 kPa of diatomic hydrogen equivalent to the solute in 100 g of metal.

The pressure-dependence of the quantities of hydrogen absorbed can be used to confirm that hydrogen absorbed from the dry gas phase at or below atmospheric pressure is accommodated in true solution and not as lithium hydride. Figures 6.11 and 6.12 are plots of the values given in Table 6.14 for binary alloys as functions of the square root of hydrogen pressure at 600°C (873 K) and 300°C (573 K) respectively, representing solutions formed at temperatures above and below the critical temperatures. Figure 6.13 is a similar plot for Alloy AA 2090 at 300°C. All of the plots are linear and extrapolate to the origin. These characteristics are consistent with Sieverts' isotherm for dissociation of a diatomic gas yielding a monatomic solute as described in Section 4.1.1.2, Chapter 4.

Compliance with Sieverts' isotherm is inconsistent with the conversion of hydrogen to a hydride phase which is characterised by a pressure-invariant region of the isotherm corresponding to equilibrium with the hydride. Moreover, because Sieverts' isotherm

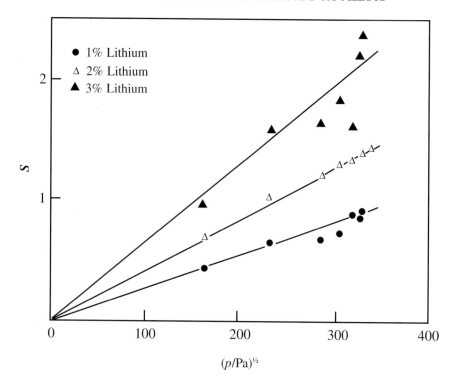

Fig. 6.12 Solubilities of hydrogen, S, in solid binary aluminium-lithium alloys as functions of pressure, p. Sieverts' isotherms for equilibrium with the gas phase at 573 K (300°C). (Anyalebechi *et al.*).[6]
$S = V/cm^3$, where V is the volume at 273 K and 101 kPa of diatomic hydrogen equivalent to the solute in 100 g of metal.

applies both above and below the critical temperatures, the discontinuities in the Vant'Hoff isobars cannot be explained by a change in the monatomic form of the solute and must be attributed to characteristics of the metal. Although the low hydrogen activities used in the solubility measurements were insufficient to nucleate lithium hydride, an internal hydride phase can be produced in the alloys by hydrogen generated at high activities in reactions of hydrated oxidation products with the metal surface, as explained in Section 8.4.3, Chapter 8.

McCracken *et al.*[23, 24] measured the solubility of the diatomic hydrogen isotope, deuterium, $^2H_2$, in the solid aluminium–2 mass% lithium alloy by the absorption-quench-desorption method to provide information for neutron scattering experiments described in Section 6.3.3.4. The values are given in Table 6.17 and plotted as the isobar in Figure 6.14, with corresponding values for natural hydrogen taken from Table 6.12 and

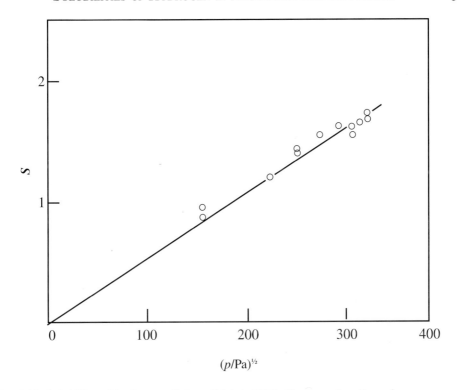

Fig. 6.13 Solubility of hydrogen, S, in solid AA 2090 alloy as a function of pressure, p. Sieverts' isotherm for equilibrium with the gas phase at 573 K (300°C). (Sargent).[11] $S = V/cm^3$, where V is the volume at 273 K and 101 kPa of diatomic hydrogen equivalent to the solute in 100 g of metal.

Figure 6.9. The solubility of deuterium is described by the pair of equations 6.38 and 6.39:

$$523 < T/K < 740: \quad \log S - \tfrac{1}{2}\log\left(\frac{p}{p^{\ominus}}\right) = \frac{-394}{T/K} + 0.417 \tag{6.38}$$

$$740 < T/K < 823: \quad \log S - \tfrac{1}{2}\log\left(\frac{p}{p^{\ominus}}\right) = \frac{-680}{T/K} + 0.617 \tag{6.39}$$

The solubility of deuterium, $^2H_2$, in the alloy is about half that of natural hydrogen, $^1H_2$, but its qualitative nature is the same. In particular it is much higher than the solubility of natural hydrogen in solid pure aluminium and it is described by a two-part isobar with a discontinuity at the same critical temperature, 740°C, as for solutions of natural hydrogen

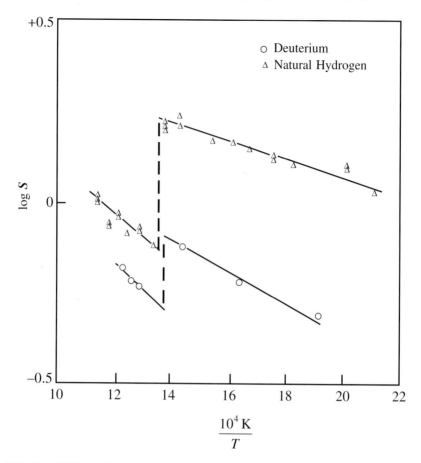

Fig. 6.14 Solubilities of deuterium, S, in solid aluminium–2 mass% lithium alloy as a function of temperature, T. Van't Hoff isobars for equilibrium with the gas phase at a pressure of 101 kPa (1 atm). Corresponding isobar for hydrogen reproduced from Figure 6.9 for comparison (McCracken *et al.*).[23, 24] $S = V/cm^3$, where V is the volume at 273 K and 101 kPa of diatomic hydrogen equivalent to the solute in 100 g of metal.

in the same alloy. The lower solubility of deuterium is attributable primarily to the effect of the larger nuclear mass on the change in the partition function for dissolution of the diatomic gas described in Section 4.1.2.2, Chapter 4.

Validity of Measured Values for the Solid Alloys
The measured values satisfy the following criteria:
Reliability of experimental methods - All of the values were measured by the absorption quench-desorption method that is preferable to isothermal methods for solid metal, as

Table 6.17 Solubilities of Deuterium, $^2H_2$, and Hydrogen, $^1H_2$, in Solid Aluminium–2 mass% Lithium Alloy at 101 kPa Pressure

Temperature		Solubility, (cm³/100 g)		Solute Mole Fraction $x_H \times 10^5$	
°C	T/K	Deuterium*	Hydrogen†	Deuterium	Hydrogen
250	523	0.49	1.19	1.1	2.9
325	598	0.59, 0.61	1.38	1.4, 1.5	3.3
420	693	0.75	1.66	1.8	4.0
500	773	0.58	0.82	1.4	2.0
520	793	0.60	0.81	1.5	2.0
550	823	0.66	0.89	1.6	2.1

*McCracken *et al.*[23, 24]   †Mean values extracted from Table 6.12

explained in Section 6.1.2. and the detailed descriptions given by the authors[10, 11, 23] again show that the equipment and procedures were designed in accordance with best practices critically reviewed in Section 6.1.2.

Reproducibility of results - Replicate values given in Tables 6.12 to 6.14 agree reasonably well although there are insufficient results for statistical analysis.

6.3.3.4 Association of Solute Hydrogen with Lithium-Rich Clusters

Binary aluminium–lithium alloys are precipitation hardening based on the metastable phase, δ'(Al$_3$Li).[26] In view of the affinity of hydrogen for lithium, McCracken *et al.*[23, 24] conducted experiments to discover if it is associated with the precipitate by comparing the ageing processes in metal with and without hydrogen, using small angle neutron scattering.

Basis of Small Angle Neutron Scattering (SANS)
Small angle neutron scattering can be used to characterise the size and dispersion of precipitate particles in ageing systems. A narrow collimated beam of monochromatic thermal neutrons is transmitted through a sample of a material under examination, selecting a neutron wavelength to suit the expected *d* spacing of the particles. Neutrons are scattered symmetrically within a cone concentric with the beam axis and are recorded by a two dimensional position sensitive detector normal to the axis. The output is an annular spectrum of intensities associated with interference effects from neutrons scattered from the assemblies of heterogenieties. The spectrum is radially averaged, corrected for artifacts

Table 6.18 Samples of Aluminium–2 mass% Lithium Alloy Prepared for Examination by Small Angle Neutron Scattering

Sample	Heat Treatment				Deuterium Content (cm³/100 g)
	Environment	Temperature		Time/h	
		°C	T/K		
1	Helium Gas	520	793	8	0
2	Deuterium Gas	520	793	50	0.60
3	Deuterium Gas	420	693	50	0.75

McCracken *et al.*[23, 24]

and standardised by suitable reference materials. A corrected spectrum gives the intensity of scattered neutrons, I, as a function of a scattering vector, Q, defined by:

$$Q = \frac{4\pi \sin\theta}{\lambda} \tag{6.40}$$

where θ is the angular displacement from the beam axis and λ is the neutron wavelength. Analysis of this function yields the required information for the system.

Use of SANS to Assess Effects of Hydrogen on Ageing an Aluminium–Lithium Alloy

McCracken *et al.*[23, 24] applied the small angle neutron scattering technique to compare the ripening of lithium clusters in an aluminium–2 mass% lithium alloy with and without dissolved hydrogen, using the isotope, $^2H_2$ (deuterium) instead of natural hydrogen, mainly $^1H_2$, to exploit its superior neutron scattering cross-section.

The deuterium was absorbed in samples of the alloy by the absorption-quench desorption method, using the sample preparation and procedures described for solubility measurements in Section 6.1.2.2 but replacing hydrogen with deuterium. Samples with different deuterium contents were produced by heating them in deuterium at different temperatures, one at 693 K and the other at 793 K for 50 hours, a time sufficient to saturate the metal as assessed from desorption rates observed in solubility determinations described in Section 6.3.3.3.[23, 24] These treatments simultaneously fulfilled the function of solution-treating the metal for subsequent ageing to yield δ'. The samples were stored under liquid nitrogen to suppress premature ageing. Deuterium-free samples for reference were given similar heat-treatments in helium which is insoluble and is a good thermal match for deuterium as a quenchant. Table 6.18 summarises these conditions.

The experiments were conducted in the neutron scattering facility at the Risø National Laboratory, Denmark. The ripening of clusters and precipitation of δ' within the samples was initiated and sustained by maintaining them at a temperature of 95°C (368 K) within

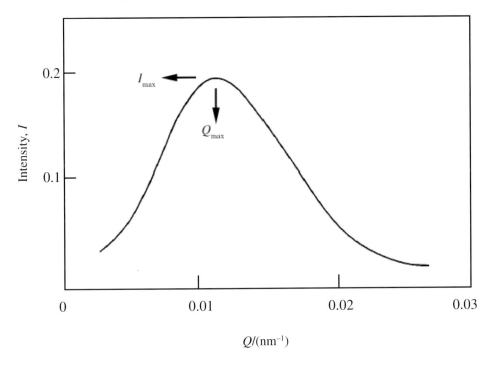

Fig. 6.15 Example of radially averaged small angle neutron scattering spectrum, giving neutron intensity, I, as a function of scattering vector, Q.
Deuterium-free aluminium–2 mass% lithium alloy solution-treated at 520°C (793 K) quenched in helium and aged for 2.6×10^5 s at 95°C (368 K). Values, I_{max} and Q_{max}, at the peak yield information on precipitates.

the sample chamber where they were heated by a small internal electric furnace. Successive spectra were recorded to monitor the progress of ageing.

Figure 6.15 gives a typical example of one from a series of spectra obtained. The neutron intensity, I is plotted as a function of momentum transfer, Q, defined by eqn 6.40. The significant feature is a peak which is associated with interference effects from neutrons scattered from heterogenieties or particles with a characteristic separation.[27] This is the expected peak following the development of the δ'precipitate and its lithium-rich precursors. No other peak was found in any of the spectra that would disclose the presence of any other fine scale entity such as hydride particles.

The significant features of a spectrum are the co-ordinates at its peak, I_{max} and Q_{max}, that are respectively measures of the size and separation of the heterogenieties responsible for scattering. In successive spectra recorded as the ageing process develops, I_{max} is

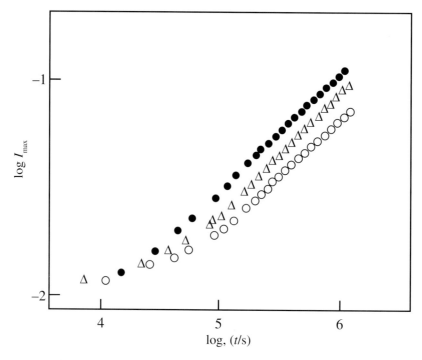

Fig. 6.16 Effect of deuterium, ^2H on the ageing of an aluminium–2 mass% lithium alloy at 95°C (368 K). Intensity of scattered neutrons I_{max}, at peak maxima in small angle neutron scattering spectra as a function of time, t.

●   Deuterium free sample, quenched from 520°C (793 K).
Δ   Sample equilibrated with deuterium at 520°C (793 K) and quenched.
○   Sample equilibrated with deuterium at 420°C (693 K) and quenched.

expected to increase as the clusters or precipitate particles grow and Q_{max} is expected to diminish as the separation between them becomes greater.

The values of I_{max} and Q_{max} for samples with and without deuterium are plotted as functions of time for comparison in Figures 6.16 and 6.17. These figures show that deuterium introduced into the alloy intervenes in the ageing process in two respects:

1. It slightly enlarges the clusters present in the alloy as quenched.
2. It delays their initial growth during ageing.
3. Raising the deuterium content enhances the effects.

These observations establish the suspected association of the solute with the clusters. As shown earlier, the isotopic effect does not modify the qualitative nature of the solutions so that the same kind of association can be assumed for natural hydrogen, $^1H_2$.

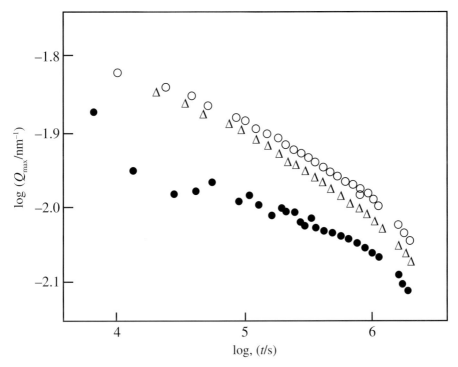

Fig. 6.17 Effect of deuterium, ^2H on the ageing of an aluminium–2 mass% lithium alloy at 95°C (368 K). Scattering vector Q_{max}, at peak maxima in small angle neutron scattering spectra as a function of time, t.
●   Deuterium free sample, quenched from 520°C (793 K).
Δ   Sample equilibrated with deuterium at 520°C (793 K) and quenched.
○   Sample equilibrated with deuterium at 420°C (693 K) and quenched.

6.3.3.5 Nature of Hydrogen Solutions in Solid Aluminium Alloys Containing Lithium
The phase diagram[25] given in Figure 6.18 shows that binary aluminium–lithium alloys with lithium contents below 2 mass% remain FCC α phase solid solutions throughout most of the temperature range to which the solubility values apply and yet the quantities of hydrogen that they occlude from the gas phase are one or two orders of magnitude greater than for the pure metal.[2, 3] The implications of the Sieverts' absorption isotherms, Van't Hoff absorption isobars and effects of hydrogen on the ageing kinetics can explain the enhanced solubilities in terms of the crystallographic features of the phase.

The Monatomic State of Hydrogen Dissolved in Aluminium–Lithium Alloys
The Sieverts' isotherms plotted in Figures 6.11 to 6.13 confirm that the hydrogen is present in the metal as atoms. They are all linear, extrapolate to the origin and show no sign of a

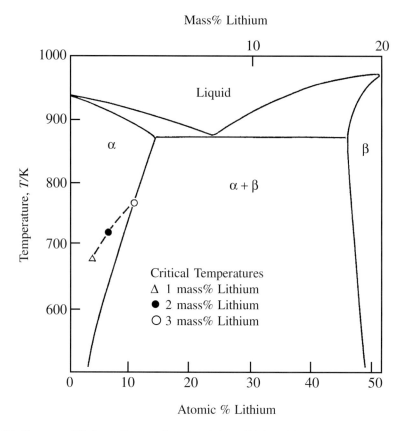

Fig. 6.18 Phase equilibrium diagram for aluminium–lithium binary system (McAlister).[25] Critical temperatures for hydrogen solutions in the α phase are superimposed.

pressure-invariant plateau that would indicate separation of a hydride phase. Moreover the equilibrium atomic ratios of hydrogen to lithium are very small and neutron scattering fails to detect any dispersion other than that expected for the ripening of δ' clusters These features are uncharacteristic of a system exhibiting hydride precipitation but consistent with the dissolution of diatomic hydrogen yielding solute atoms, so that the effect of lithium is to enhance the solubility of hydrogen in the α phase. Thus in view of the discussion on the stability of lithium hydride in Chapter 4, Section 4.3.3.1, solutions of hydrogen in equilibrium with the gas phase for $p \leq 101$ kPa, are apparently metastable with respect to an internal hydride phase.

The subcutaneous hydride particles that are sometimes observed in manufactured products are produced by very high hydrogen activities generated by reaction of the metal with surface hydrates or humid atmospheres due to industrial malpractice as described in Chapter 9.

Table 6.19 Solubility of Hydrogen in Aluminium–Lithium Binary Alloys at 101 kPa - Conversion of Hydrogen Content, cm³/100 g[†] to Mole Fraction, x_H

Temperature		1 Mass% Lithium		2 Mass% Lithium	
°C	10^4 K/T	cm³/100 g[†]	x_H	cm³/100 g[†]	x_H
200	21.1	0.67	1.61×10^{-5}	1.03	2.48×10^{-5}
225	20.1	0.60	1.45×10^{-5}	1.16	2.80×10^{-5}
250	19.1	0.76	1.83×10^{-5}	1.19	2.87×10^{-5}
275	18.2	0.83	2.00×10^{-5}	1.25	3.01×10^{-5}
300	17.5	0.89	2.14×10^{-5}	1.30	3.13×10^{-5}
325	16.7	0.97	2.34×10^{-5}	1.38	3.33×10^{-5}
350	16.1	1.00	2.41×10^{-5}	1.38	3.33×10^{-5}
375	15.4	1.05	2.53×10^{-5}	1.45	3.49×10^{-5}
400	14.86	0.97	2.34×10^{-5}	1.46	3.52×10^{-5}
425	14.33	0.61	1.47×10^{-5}	1.66	4.00×10^{-5}
450	13.83	0.60	1.45×10^{-5}	1.64	3.95×10^{-5}
475	13.37	-	-	0.72	1.74×10^{-5}
500	12.94	0.69	1.66×10^{-5}	0.82	1.98×10^{-5}
525	12.53	0.71	1.71×10^{-5}	0.81	1.95×10^{-5}
550	12.15	0.79	1.90×10^{-5}	0.89	2.14×10^{-5}
575	11.79	0.81	1.95×10^{-5}	0.89	2.14×10^{-5}
600	11.46	0.84	2.02×10^{-5}	1.02	2.46×10^{-5}

[†]Mean values extracted from Table 6.12

Discrete Preferred Sites for Hydrogen Dissolved in Aluminium–Lithium Alloys

The application of Van't Hoff isobars to identify the crystallographic sites that solute hydrogen atoms occupy in pure FCC metals, explained in Section 4.1.2.3, can be extended to alloys that remain FCC throughout the temperature range considered. This includes the 1 and 2 mass% lithium binary alloys which provide sufficient information to establish the role of lithium in generating additional sites for hydrogen. It cannot be further extended to more complex alloys including aluminium–3 mass% lithium, AA 2090, AA 8090 and AA 8091 in which other phases or their precursors intervene.

To correlate the information from the isobars with crystallographic structures, the solubilities given for the 1 and 2 mass% lithium alloys as hydrogen contents in Table 6.12 are converted to mole fractions of hydrogen, x_H, in Table 6.19 using eqn 6.3 and plotted as isobars in Figure 6.19. These isobars acquire significance when compared with the

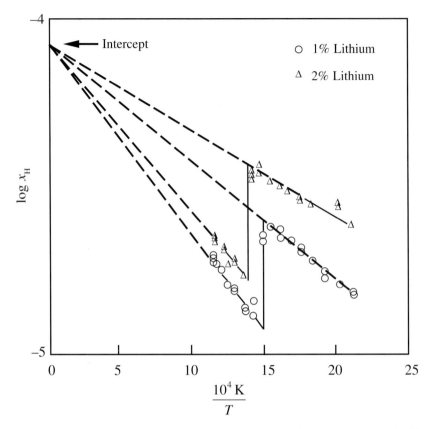

Fig. 6.19 Van't Hoff isobars at 101 kPa (1 atm) for hydrogen solutions in binary aluminium–lithium alloys with solute activity referred to the infinitely dilute mole fraction standard state, $a_H \rightarrow x_H$ as $x_H \rightarrow 0$. All of the isobars converge to a common intercept at $K/T = 0$ with the value $[\log x_H]_{K/T=0} = -4.07$.

corresponding isobar for pure aluminium, given in Figure 4.2. Two essential differences indicate the sites of the solute and identify the reason for the high solubilities.

1. Entropies of Mixing

The isobar for pure aluminium, given in Figure 4.2 intercepts the abscissa at $\log x_H = -2.40$, which includes the standard entropy of mixing for solute atoms in octahedral interstices FCC lattices. In contrast, the isobars for the alloys, given in Figure 6.19 intercept the abscissa at a much smaller value, $\log x_H = -4.07$. Since the entropies for dissociation of the gas and for change in partition function are the same, the lower value implies that the entropies of mixing for the solute are much lower, showing that the increased quantity of hydrogen absorbed on introducing lithium as an alloy component is accommodated in fewer sites. From the values of the intercepts, the ratio of the number of these sites to the number of octahedral interstices is 0.02.

2. Binding Enhalpies

Since the introduction of lithium raises the capacity of the metal for hydrogen but introduces relatively few extra sites to accommodate it, it follows that the new sites are more densely populated with solute atoms than the existing interstitial sites. This in turn implies that the binding enthalpy at the new sites is stronger so that the overall process in which hydrogen dissociates and dissolves as atoms is expected to be less strongly endothermic for the alloys than for pure aluminium. This is manifest by the reduced slopes of the isobars for the alloys compared with the isobar for pure aluminium given in Figure 4.2.

All of this evidence establishes that the enhanced solubilities are due to the provision of discrete preferred sites that are available to the solute. These sites are identified as the precursors of the δ' precipitate from their association with the solute manifest by delayed ripening revealed by small angle neutron scattering analysis.

The regular octahedral interstitial sites in the alloys contribute a quantity of hydrogen of the same order of magnitude as in pure aluminium so that it is only a small fraction of the total. However, the partition of the solute between the two kinds of site is essential because most of the solute is isolated in the preferred sites and the interstitial sites provide the only continuous path through which it can diffuse. The system is an example of diffusion in a field of traps and the mathematical model applied to it in Sections 7.4.1 and 7.4.2, Chapter 7 further supports the concept of discrete sites for hydrogen.

The Significance of the Critical Temperatures

The partial isobars above and below the critical temperature converge to the same intercept, indicating equal entropy of solution, so that the number of sites for hydrogen is the same above and below the critical temperature. The rise in solubility at the discontinuity introduces a different slope for lower temperatures, signifying an abrupt increase in solute binding enthalpy, perhaps due to reordering on an atomic scale within the clusters. The sensitivity of the electrical resistivity of aluminium–lithium alloys to quenching from above or below a comparable threshold temperature[28, 29] may be a related effect.

Effect of Lithium Content

The intercept is the same for both the 1% and 2 mass% lithium alloys indicating that the numbers of sites for hydrogen is insensitive to the composition of the α phase within this range but the solute binding energy in the alloy with the higher lithium content is greater and the critical temperature is higher.

Location of Dissolved Hydrogen in Multiphase Alloys

Multiphase alloys contain other species that could provide discrete preferred hydrogen sites, including traces of the β phase and precursors of the T_1 (Al_2CuLi) and T_2 (Al_6CuLi_3) phases in the aluminium–magnesium–copper–lithium system. Such systems are obviously too complex for rigorous analysis of their hydrogen absorption isobars. However the Van't Hoff format as used for the 3 mass% lithium alloy, AA 2090, AA 8090 and AA 8091 alloys in Figures 6.9 and 6.10 have empirical utility in relating enhancement of hydrogen solubility to aspects of their phase equilibria. For example, the wayward form of the isobar for the 3 mass% lithium alloy in Figure 6.9 is probably due to its duplex structure in most of the temperature range considered.

6.4 REFERENCES

1. A. Sieverts, *Z. Metallkunde*, **21**, 1929, 37–45.
2. C.E. Ransley and H. Neufeld, *J. Inst. Metals*, **74**, 1948, 599.
3. W.R. Opie and N.J. Grant, *Trans. AIME*, **188**, 1950, 1237.
4. R. Eborall and C.E. Ransley, *J. Inst. Metals*, **71**, 1945, 525.
5. D.E.J. Talbot and D.A. Granger, *J. Inst. Metals*, **92**, 1963–4, 290.
6. P.N. Anyalebechi, D.E.J. Talbot and D.A. Granger, *Met. Trans.*, **20B**, 1989, 523.
7. W. Eichenauer and A.Pebler, *Z. Metallkunde*, **48**, 1957, 373.
8. W. Eichenauer et al., *Z. Metallkunde*, **59**, 1968, 613.
9. W. Eichenauer H. Kunzig and A.Pebler, *Z. Metallkunde*, **49**, 1958, 220–225.
10. P.N. Anyalebechi, D.E.J. Talbot and D.A. Granger, *Met. Trans.*, **19B**, 1988, 227.
11. M.A. Sargent, "The Solubility of Hydrogen in Some Commercial Aluminium–Lithium Alloys", Ph.D. Thesis, Brunel University, 1988.
12. D.E.J. Talbot and P.N. Anyalebechi, *Mat. Sci. and Tech.*, **4**, 1988, 1.
13. W. Eichenauer, K. Hattenbach and A. Pebler, *Z. Metallkunde*, **52**, 1961, 682.
14. C.E. Ransley, D.E.J. Talbot and H. Barlow, *J. Inst. Metals*, **86**, 1957–58, 212 and British Patent No.684865.
15. D.A. Granger, "Telegas for Determining Hydrogen in the Foundry Industry" *Proc. International Molten Aluminum Symposium*, City of Industry, California, 1986, 417–431.
16. J-P. Martin, F. Tremblay and G. Dubé, *Light Metals*, 1989, 903.
17. P.D. Hess, *J. Met.*, 1973.
18. P. Röntgen and F. Möller, *Metallwirt., Metallwiss., Metalltech.*, **13**, 1934, 81 and 97.
19. M. Weinstein and J.F. Elliot, *Trans Met. Soc. AIMME*, **227**, 1963, 382.
20. L. Luckmeyer-Hasse and H. Schenck, *Arch. Eisenhutt. Wes.*, **6**, 1932–3, 209.
21. H. Schenck and K.W. Lange, *Arch. Eisenhutt. Wes.*, **37**, 1966, 739.
22. J. Koenman and A.J. Metcalf, *Trans. Amer. Soc. Metals*, **51**, 1959, 1072.
23. C.G. McCracken, "The Intrinsic and Extrinsic Solubility of Hydrogen in Aluminium Lithium Based Alloys", Ph.D., Thesis, Brunel University, 1994.
24. C.G. McCracken, D.E.J. Talbot and J. Skov Pedersen, *Proc. Int. Conf. Microstructures and Mechanical Properties of Aging Materials*, Chicago 1992, TMS 1993, 481–487.
25. A.J. McAlister, *Bulletin of Alloy Phase Diagrams*, **3**, 1982, 177.

26. J.M. Silcock, *J. Inst. Metals*, **88**, 1959–60, 357.
27. E. Caponetti, E.M.D'Aguanno, R. Triolo and S. Spooner, *Phil. Mag. B*, **63**, 1991, 1201.
28. S. Ceresara, A. Giarda and A. Sanchez, *Acta. Met.*, **34**, 1986, 1021.
29. R. Kamel, A.R. Ali and Z. Farid, *Phil. Mag.*, **35**, 1977, 97.

7. The Diffusion of Hydrogen in Aluminium and its Alloys

Diffusion is the thermally activated transport of matter within a host medium by random spontaneous jumps of fundamental particles between adjacent sites within the host, impartially exploring the microscopic complexions of the system, progressively reducing existing activity gradients of the diffusing species. Diffusion rates depend *inter alia* on a correlation factor, ≤ 1, expressing the probability that adjacent sites for migrating atoms are available. For the dilute interstitial solutions of hydrogen in aluminium, all sites adjacent to any particular hydrogen atom are almost certainly empty, so that the correlation factor is virtually unity. Moreover the activation energy to transfer such a small atom between adjacent sites is low. Hence the diffusion is expected to be faster than for substitution solutes which have large atoms and low correlation factors.

7.1 MATHEMATICS OF DIFFUSION

The mathematics of diffusion is covered in Crank's classic monograph[1] and texts by Barrer,[2] Jost[3] and Carslaw and Jaeger.[4] In the present context, it is sufficient to give some equations for uniform isotropic media and to show how to modify them to describe diffusion in fields of traps, simulating the mobility of hydrogen in manufactured aluminium products.

7.1.1 DIFFUSION IN A UNIFORM ISOTROPIC MEDIUM

The theory is based on Fick's[5] adaptation of Fourier's equations for heat conduction[6] to describe diffusion by replacing thermal parameters with their concentration counterparts. The basic principle is that the flux of a substance through a unit section of a uniform medium in which it diffuses is proportional to its concentration gradient normal to the section, i.e:

$$J_x = -D\frac{\partial c}{\partial x} \tag{7.1}$$

where: J_x is the flux at the co-ordinate, x.
          c is the instantaneous solute concentration at x.
          D is called the *diffusion coefficient.*

Eqn 7.1 describes the flux in terms of concentration gradients but diffusion is strictly driven by activity gradients, so that D is constant only for solutes exhibiting Raoultian or Henrian activities. Dilute solutions of hydrogen in aluminium are Henrian so that an assumption that D is constant is justified.

Arrhenius' relation gives the temperature-dependence of D for a restricted range:

$$D = D_o \exp \frac{\Delta G^*}{RT} \qquad (7.2)$$

where the factor, D_o and the activation energy, G^*, are system properties.

Considering the mass balance at a point defined by rectangular co-ordinates (x, y, z) yields the differential equation:

$$\frac{\partial c}{\partial t} + \frac{\partial J_x}{\partial x} + \frac{\partial J_y}{\partial y} + \frac{\partial J_z}{\partial z} = 0 \qquad (7.3)$$

For isotropic media with constant D, substituting for J_x, J_y and J_z from eqn 7.1 yields:

$$\frac{\partial c}{\partial t} = D \left[\frac{\partial^2 c}{\partial x^2} + \frac{\partial^2 c}{\partial y^2} + \frac{\partial^2 c}{\partial z^2} \right] \qquad (7.4)$$

Eqn 7.4 can be adapted to other geometries, including semi-infinite media, sheets, cylinders and spheres. Crank,[1] Barrer,[2] Jost[3] and Carslaw and Jaeger[4] give solutions for these geometries with conditions including non uniform initial conditions, variable boundary conditions, anisotropic media, diffusion in fields of traps, and variable diffusion coefficients.

Two geometries are particularly relevant to wrought metal products, an infinite plane sheet representing flat rolled products and an infinite cylinder, representing rolled or extruded round rod. Differential equations for these geometries and solutions for the absorption of hydrogen from a surface source or its desorption to a surface sink are given in Sections 7.1.1.1 and 7.1.1.2 assuming some simple initial and boundary conditions.

7.1.1.1 Absorption in or Desorption from an Infinite Plane Sheet
Diffusion is normal to the surface, the initial concentration of the diffusing species is uniform, the surface concentration is constant and the diffusion coefficient is constant.

Differential Equation and Conditions:
Infinite sheet, $2a$ thick with transverse co-ordinate, $\pm x$, referred to the midplane. For diffusion restricted to the x co-ordinate, eqn 7.4 reduces to:

$$\frac{\partial c}{\partial t} = D \frac{\partial^2 c}{\partial x^2} \tag{7.5}$$

Initial condition: $c = c_o$, $-a < x < +a$, $t = 0$.

Boundary condition: $c = c_s$, $x = \pm a, t > 0$

where c_o is the uniform initial concentration and c_s is the constant surface concentration.

Solutions to the Equation Satisfying the Conditions:

1. A solution in the form of a trigonometrical series yielding concentration profiles through the thickness of the sheet is:[1]

$$\frac{c_x - c_o}{c_s - c_o} = 1 - \frac{4}{\pi} \sum_{n=0}^{\infty} \frac{(-1)^n}{2n+1} \exp \frac{-D(2n+1)^2 \pi^2 t}{4a^2} \cos \frac{(2n+1)\pi x}{2a} \tag{7.6}$$

where c_x is the concentration at a distance, $\pm x$, from the mid plane after time, t.

2. A solution yielding the fraction of solute absorbed or desorbed after time, t is:[1]

$$\frac{Q_t}{Q_\infty} = 1 - \frac{8}{\pi^2} \sum_{n=0}^{\infty} \frac{1}{(2n+1)^2} \exp \frac{-D(2n+1)^2 \pi^2 t}{4a^2} \tag{7.7}$$

where: Q_t is the quantity of substance that has entered or left the plate in time, t and Q_∞ is the corresponding quantity after infinite time.

7.1.1.2 Absorption in or Desorption from an Infinite Cylinder

Diffusion is radial, the initial concentration of the diffusing species is uniform, the surface concentration is constant and the diffusion coefficient is constant.

Differential Equation and Conditions:

Cylinder of radius, a, with radial co-ordinate, r, referred to the axis. Transposing eqn 7.4 to cylindrical co-ordinates for diffusion restricted to the radial co-ordinate:[1]

$$\frac{\partial c}{\partial t} = \frac{1}{r} \frac{\partial}{\partial r} \left[rD \frac{\partial c}{\partial r} \right] \tag{7.8}$$

Initial condition: $c = c_o$, $0 < r < a$, $t = 0$.

Boundary condition: $c = c_s$, $x = r$, $t > 0$.

where: c_o is the uniform initial concentration and c_s is the constant surface concentration.

Solutions to the Equation Satisfying the Conditions:

1. A solution in the form of Bessel functions yielding radial concentration profiles is:[1]

$$\frac{c_r - c_o}{c_s - c_o} = 1 - 2a \sum_{n=1}^{\infty} \frac{J_o(\beta_n r/a)}{\beta_n^2 J_1(\beta_n)} \exp \frac{-Dt\beta_n^2}{a^2} \tag{7.9}$$

where: c_r is the concentration at a distance, r, from the axis of the cylinder after time, t, $J_o(\beta_n)$ and $J_1(\beta_n)$ are the Bessel functions of the first kind of zero and first orders respectively; β_n is the nth root of the equation, $J_o(\beta_n) = 0$.

2. A solution yielding the fraction of solute absorbed or desorbed after time, t, is:[1]

$$\frac{Q_t}{Q_\infty} = 1 - \sum_{n=1}^{\infty} \frac{4}{\beta^2} \exp \frac{-Dt\beta_n^2}{a^2} \tag{7.10}$$

7.1.1.3 Dimensionless Parameters

Conventions and Purpose
It is convenient to give general solutions to common diffusion problems in terms of dimensionless parameters. Time, t, is incorporated in the parameter, Dt/a^2, concentrations, c are described as ratios to standard concentrations, e.g. an initial concentration, c_o or a surface concentration, c_s, and spacial co-ordinates are described by ratios to dimensions, a, characteristic of the geometric forms in which diffusion proceeds, e.g. x/a or r/a for linear or radial co-ordinates respectively. The solutions are matched to particular problems by inserting values for the physical quantities, c_o, c_s, D, t, and a into the dimensionless parameters. Crank[1] and Carslaw and Jaeger[4] give comprehensive information in this form.

Applications to Infinite Plane Sheet and Infinite Cylinder
It is useful and instructive to display the development of concentration profiles and the progress of absorption or desorption by graphs plotted with dimensionless parameters as co-ordinates, as illustrated in Figures 7.1 to 7.3:

1. Figures 7.1 and 7.2 show concentration profiles given by eqns 7.6 and 7.9.
2. Figure 7.3 shows fractions of solute, Q_t/Q_∞, entering or leaving an infinite plate or infinite cylinder as functions of Dt/a^2, derived from eqns 7.7 and 7.10.

   For quantitative applications it is usually more convenient to tabulate numerical relations between the dimensionless parameters evaluated from eqns 7.6, 7.7, 7.9 and 7.10. Table 7.1 gives values of Dt/a^2 for selected values of Q_t/Q_∞ for an infinite plate and an infinite cylinder for later reference.

7.1.2 Diffusion in a Field of Traps

The description of diffusion given in Section 7.1.1 applies to idealised media free from preferred sites for the diffusing substance. It does not describe the transport of hydrogen through manufactured aluminium and aluminium alloy products that invariably contain

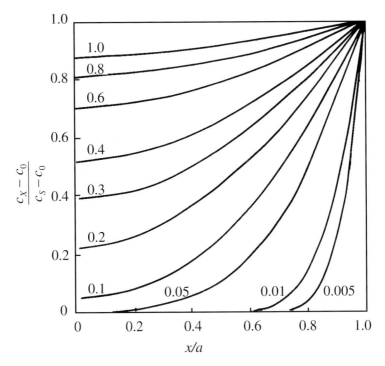

Fig. 7.1 Diffusion into an infinite sheet. Successive concentration profiles through the thickness of a sheet with an initial uniform concentration, constant surface concentration and constant diffusion coefficient. Calculated from eqn 7.6 for the values of the dimensionless parameter, Dt/a^2 appended to the profiles.

$\pm a$ is the half-thickness of the sheet measured from its midplane.

c_x is the concentration at a distance, $\pm x$, from the midplane after time, t.

c_o is the initial concentration.

c_s is the surface concentration.

D is the diffusion coefficient.

Table 7.1 Values of the Parameter Dt/a^2 in eqns 7.7 and 7.10 Corresponding to Fractions of Solute, Q_t/Q_∞, Diffusing in or out of an Infinite Plane Sheet or an Infinite Cylinder after Elapsed Time, t[†]

Infinite Plane Sheet, Thickness, 2a								
Q_t/Q_∞	0.40	0.50	0.60	0.70	0.80	0.90	0.95	
Dt/a^2	0.12	0.19	0.28	0.40	0.56	0.82	1.06	
Infinite Cylinder, Radius a								
Q_t/Q_∞	0.40	0.50	0.60	0.70	0.80	0.90	0.95	0.98
Dt/a^2	0.041	0.065	0.097	0.144	0.212	0.335	0.460	0.630

[†]Carslaw and Jaeger[4]

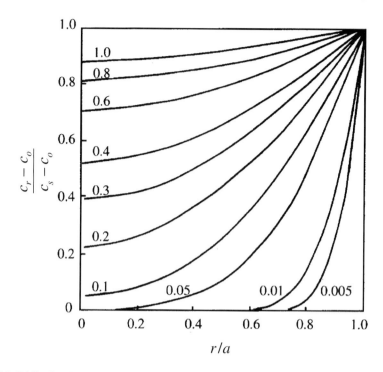

Fig. 7.2 Diffusion into an infinite cylinder. Successive radial concentration profiles through a cylinder with an initial uniform concentration, constant surface concentration and constant diffusion coefficient. Calculated from eqn 7.6 for the values of the dimensionless parameter, Dt/a^2 appended to the profiles.

    a is the radius of the cylinder.
    c_r is the concentration at a radius, r, from the axis of the cylinder after time, t.
    c_o is the initial concentration.
    c_s is the surface concentration.
    D is the diffusion coefficient.

hydrogen traps of the various kinds described in Section 4.3. This must be treated as a problem in which a fraction of the diffusing substance is immobilised by reversible trapping[1] at sites distributed throughout the medium and diffusion equations are transformed to allow for it. For diffusion in one dimension eqn 7.5 becomes:

$$\frac{\partial c}{\partial t} = D\frac{\partial^2 c}{\partial x^2} - \frac{\partial c_i}{\partial t} \qquad (7.11)$$

and for radial diffusion in a cylinder, eqn 7.8 becomes:

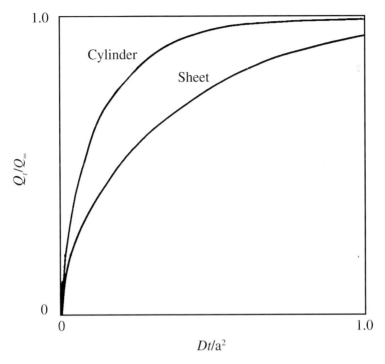

Fig. 7.3 Fraction of solute, Q_t/Q_∞, diffusing into or out of an infinite sheet or an infinite cylinder with uniform initial concentration and constant surface concentration, as functions of the dimensionless parameter, Dt/a^2 in eqn 7.7 and 7.10.
where: D is the diffusion coefficient, t is the elapsed time.
        a is the half-thickness of the sheet or the radius of the cylinder.

$$\frac{\partial c}{\partial t} = \frac{1}{r}\frac{\partial}{\partial r}\left[rD\frac{\partial c}{\partial r}\right] - \frac{\partial c_i}{\partial t} \qquad (7.12)$$

where c_i is the concentration of immobilised substance.

To apply eqns 7.11 and 7.12 the relation between c_i and c must be given. Two relations are relevant to the transport of hydrogen in aluminium products, a *linear* isotherm describing the formation of *monoatomic* hydrogen traps from the solute and a *non linear* isotherm describing the formation of *diatomic* traps. In applying either of these isotherms it is assumed that the diffusing system is under diffusion control, so that local equilibrium is maintained between the free and trapped components of the diffusing substance.

7.1.2.1 Linear Isotherm

The concentration of the immobilised component, c_i, is proportional to the concentration of the free component, c:

$$c_i = K c \tag{7.13}$$

Substituting for c_i in eqn 7.11:

$$\frac{\partial c}{\partial t} = D \frac{\partial^2 c}{\partial x^2} - K \frac{\partial c}{\partial t} \tag{7.14}$$

Rearranging:

$$\frac{\partial c}{\partial t} = \frac{D}{K+1} \times \frac{\partial^2 c}{\partial x^2}$$

$$= D' \frac{\partial^2 c}{\partial x^2} \tag{7.15}$$

where: K = equilibrium constant.
c = concentration of the free component.
c_i = concentration of the immobilised component.
D = diffusion coefficient.

This equation is in the form of eqn 7.5, replacing the diffusion coefficient, D, with a coefficient, $D' = D/(K + 1)$. It is convenient to use Crank's[1] term, "*effective diffusion coefficient*" to describe D', although it is not a true physical quantity and is application dependent. A corresponding equation for cylindrical co-ordinates with diffusion restricted to the radial co-ordinate can be derived from eqn 7.12:

$$\frac{\partial c}{\partial t} = \frac{1}{r} \frac{\partial}{\partial r} \left[r D' \frac{\partial c}{\partial r} \right] \tag{7.16}$$

The solutions given in eqns 7.6, 7.7, 7.9 and 7.10 and Table 7.1 remain valid when a diffusion coefficient, D is replaced by a corresponding effective diffusion coefficient, D'.

7.1.2.2 Non-Linear Isotherm

The relation between concentrations of immobilised and free components is of the form:

$$c_i = K c^n \tag{7.17}$$

Substituting for c_i in eqn 7.11:

$$\frac{\partial c}{\partial t} = D \frac{\partial^2 c}{\partial x^2} - K \frac{\partial c^n}{\partial t} \tag{7.18}$$

Eqn 7.18 is non-linear and difficult to apply but if K is large so that $\partial c/\partial t$ is negligible compared with $\partial c_i/\partial t$, a more manageable equation can be derived from the simplified form of eqn 7.11:

$$\frac{\partial c_i}{\partial t} = D \frac{\partial^2 c}{\partial x^2} = \frac{\partial}{\partial x} \left[D \frac{\partial c}{\partial x} \right] \tag{7.19}$$

Substituting for c in eqn 7.19 yields:

$$\frac{\partial c_i}{\partial t} = D \frac{\partial^2 c}{\partial x^2} = \frac{\partial}{\partial x} \left[D \frac{\partial}{\partial x} (c_i / K)^{\frac{1}{n}} \right] \tag{7.20}$$

$$= \frac{\partial}{\partial x} \left[\frac{D}{n} \left(\frac{1}{K} \right)^{\frac{1}{n}} c_i^{(1-n)/n} \frac{\partial c_i}{\partial x} \right] \tag{7.21}$$

$$= \left\{ (D/n)(1/K)^{1/n} c_i^{(1-n)/n} \right\} \frac{\partial^2 c_i}{\partial x^2} \tag{7.22}$$

Collecting terms:

$$\frac{\partial c_i}{\partial t} = D' \frac{\partial^2 c_i}{\partial x^2} \tag{7.23}$$

shows that eqn 7.23 is in a form similar to a standard diffusion equation, where the coefficient, D, is replaced by an effective coefficient, D', given by:

$$D' = \left\{ (D/n)(1/K)^{1/n} c_i^{(1-n)/n} \right\} \tag{7.24}$$

Inspection of eqn 7.24 shows that the effective coefficient, D', for diffusion in a field of non linear traps is not a unique physical property of a diffusing system but an application dependent parameter that also depends the quantity of diffusing substance, c_i. In this case, too, eqns 7.6, 7.7, 7.9 and 7.10 and Table 7.1 remain valid when an intrinsic diffusion coefficient, D is replaced by a corresponding effective diffusion coefficient, D'.

7.2 DETERMINATION OF DIFFUSIVITY COEFFICIENTS

7.2.1 APPROACH

Diffusion coefficients are determined by fitting solutions of diffusion equations to the results of corresponding practical measurements. A convenient experimental procedure is to equilibrate a sample with hydrogen gas and then to record the evolution of the absorbed hydrogen as a function of time when the sample is maintained at a prescribed temperature in an evacuated enclosure, from which the gas is pumped continuously into a low pressure system for collection and measurement. Cylindrical samples, 10 to 12 mm in diameter, are most suitable because they are easy to prepare and the measurements fit conveniently into a working day but other geometries can be used. Eqn 7.10 applies to finite cylinders without serious error if the aspect ratio ≥ 5. The initial and boundary conditions are:

$$c = c_o \ \ 0 < x < r \ \ t = 0; \qquad c = 0 \ \ x = r, \ t > 0$$

7.2.2 Procedures for the Solid Metal

Sample preparation, equipment and procedures are identical to those used for determination of solubilities by the absorption-desorption methods used to determine solubilities of hydrogen in solid metal, as described in Section 6.1.2. Indeed it is often convenient and economic to combine measurements of solubility and diffusivity, provided the following aspects of the procedures receive meticulous attention:

1. To assume infinite cylindrical geometry, the aspect ratio of the sample must be ≥ 5.
2. The sample dimensions must be accurately measured.
3. The volume of hydrogen evolved must be monitored accurately and continuously.

The desorption process must be isothermal and the quantity of hydrogen evolved must be known as a function of time *ab initio*. Desorption must be under diffusion control, a condition that can be verified from results for geometries with different surface/volume ratios.

In the absorption-quench-desorption version of the procedure, in which the sample is quenched after equilibrating with hydrogen, the main source of error is the period of non-isothermal diffusion while the sample is reheated to the prescribed temperature for the desorption stage. It is minimised using radio frequency (RF) induction heating, which can heat a 12 mm diameter cylindrical sample of aluminium to any temperature below the melting point, 660°C in < 60 s. In the alternative isothermal version in which the sample remains at the test temperature throughout absorption and desorption, the main source of error is in the correction for hydrogen lost while the gas phase is removed before hydrogen desorbed from the sample can be diverted into the collection system, as explained in Section 6.1.2.2.

To calculate D, the total hydrogen content of the sample corresponding to Q_∞ in eqn 7.10, is needed. The hydrogen cannot all be recovered at low temperatures and when enough is collected for the calculations the balance is extracted at a higher temperature.

7.2.3 Evaluation of Diffusion Coefficients

A table of values of Dt/a^2 as a function of Q_t/Q_∞, such as Table 7.1 is required. Co-ordinates, $(Q_t/Q_\infty, t)$, are extracted from any point on the experimental desorption curve and the numerical value of the parameter, Dt/a^2, corresponding to the ordinate, Q_t/Q_∞, is ascertained from the Table; let this value be X. The diffusion coefficient, D, is given by:

Table 7.2 Evaluation of a Diffusion Coefficient from a Desorption Curve. Example is for Desorption of Hydrogen from a Cylindrical Sample of an aluminium-3 mass% Lithium Alloy at 500°C (873 K)[*]

Q_t/Q_∞	0.40	0.50	0.60	0.70	0.80	0.90
Parameter[†], Dt/a^2	0.041	0.065	0.097	0.144	0.212	0.335
Elapsed Time[‡], t/s	925	1675	2775	4300	6400	9075
Cylinder Radius, a/cm	0.533	0.533	0.533	0.533	0.533	0.533
$(D/\text{cm}^2\text{s}^{-1}) \times 10^5 = $ Parameter $\times$ (a^2/t)	1.24	1.09	0.98	0.94	0.93	1.03

[†]Extracted from Table 7.1.
[‡]Corrected for non-isothermal period during heating.
[*]Anyalebechi.[7]

$$D = X \times (a^2/t)$$

where: a is the radius of the cylinder and
           t is the corresponding time co-ordinate.

If the process is under diffusion control, all pairs of co-ordinates yield similar values for D, as illustrated in Table 7.2.

7.3 DIFFUSIVITY IN ALUMINIUM

7.3.1 TRUE DIFFUSIVITY IN SOLID PURE ALUMINIUM

True diffusivities have been measured only by Eichenauer and his co-workers,[8, 9] mainly for 99.5% pure aluminium, supplemented by a few measurements on 99.994% pure metal for comparison. These authors have an outstanding reputation for reliability of their measurements on solid metals inspiring confidence in their results. The diffusivity was measured by the isothermal absorption-desorption method, described earlier in Section 7.2, using large samples for high sensitivity, i.e. 2 cm diameter × 15 cm cylinders and 5.9 cm diameter spheres. The two geometries with different surface/volume ratios were used to establish that the measurements were not influenced by surface processes. Diffusion coefficients were calculated by fitting the experimental desorption curves to appropriate functions fitted to eqn 7.10 and the corresponding equation for a sphere.[1, 4]

An essential pre-requisite was to eliminate molecular hydrogen traps present in aluminium samples used for the measurements. The samples were prepared by maintaining them at the temperatures prescribed for the measurements for periods of up to 336 hours, during which they were repeatedly subjected to cycles in which they were alternately

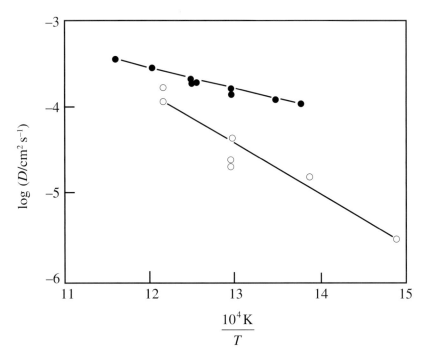

Fig. 7.4 Diffusivity of hydrogen in solid pure aluminium as functions of temperature.
● True diffusion coefficient, D, in consolidated metal free from hydrogen traps (Eichenauer and Pebler).[8]
○ Effective diffusion coefficient, D', for diffusion in the presence of diatomic hydrogen traps prevailing in manufactured products (Ransley and Talbot[10] and Talbot[16]).

equilibrated with gaseous hydrogen at a sub atmospheric pressure, 84 kPa, and degassed by desorption into an evacuated system without intermediate exposure to air. Any pores present would have collapsed under surface tension during the evacuation stages by the mechanism described in Section 4.2.3. The authors may not have appreciated how essential was their meticulous sample preparation, because the prevalence of microporosity was not generally appreciated when the measurements were made but in retrospect it was ideally suited to the purpose. The grain size is not quoted but it was probably large enough to avoid significant influence of short circuiting by grain boundaries, because the samples were derived from semi-continuously cast stock. Thus the prepared samples were probably in an ideal condition closely approaching the model of formal theory.

  The results were given as the Arrhenius plot reproduced in Figure 7.4 from which the authors extracted the following equation for the temperature dependence of diffusivity:

$$D = 0.21 \exp \frac{-5505}{T/K} \ \text{cm}^2\,\text{s}^{-1} \tag{7.25}$$

7.3.2 DIFFUSIVITY IN MANUFACTURED PRODUCTS

The quantities of hydrogen occluded in aluminium products during manufacture are invariably higher than can be accommodated in solid solution and the hydrogen is distributed between solution and a dispersed internal gas phase trapped in residues from ingot porosity described in Chapter 9 or micropores nucleated as explained in Chapter 4. The effects are well known and well documented.[10-17]

The pores constitute a field of diatomic hydrogen traps which significantly modifies diffusivity. Most of the hydrogen is in the pores and equilibrium with the mobile monoatomic solute is described by the non-linear isotherm given in eqn 7.17, where the exponent, n, equals 2:

$$c_i = K c^2 \tag{7.26}$$

suggesting the application of eqn 7.23 viz:

$$\frac{\partial c_i}{\partial t} = D' \frac{\partial^2 c_i}{\partial x^2} \tag{7.23}$$

where the effective diffusion coefficient, D', governs diffusion in the field of traps.

It is not feasible to derive D' from first principles but since the equation is in the form of a standard diffusion equation, it can be approached from empirical desorption kinetics.

Talbot[16] monitored the isothermal desorption of hydrogen from a 16 mm diameter extruded rod of 99.99% pure aluminium, with an existing uniform hydrogen content of 0.32 ± 0.02 cm³/100 g, acquired during manufacture. Cylindrical samples, 12 mm diameter × 60 mm long were dry turned from the rod to a fine finish but otherwise untreated. The temperature of every sample was raised to its prescribed value by RF induction heating and the hydrogen evolved was collected and measured continuously, using essentially the vacuum extraction method described in Section 7.2.2. The desorption curves[16] are reproduced in Figure 7.5. The effective diffusion coefficients were evaluated by matching pairs of co-ordinates extracted from the empirical desorption curves to the corresponding theoretical values of Dt/a^2 given in Table 7.1. The values obtained are given in Table 7.3.

In Table 7.3 it is evident that for every desorption curve, the same value for D' is obtained, whatever pair of co-ordinates, $(Q_t/Q_\infty, t)$ is selected to evaluate it, i.e. the effective diffusion coefficients were virtually constant as the gas was evolved, justifying the application of eqn 7.23. Interpolating the values in Table 7.3 yields D' as a function of temperature:

$$D' = 1.2 \times 10^5 \exp \frac{-16900}{T/K} \ cm^2 \, s^{-1} \tag{7.27}$$

It is common experience from innumerable measurements of hydrogen contents by vacuum extraction that D' is insensitive to the source of the metal, and before the nature of hydrogen occlusion in manufactured aluminium products was appreciated, it was

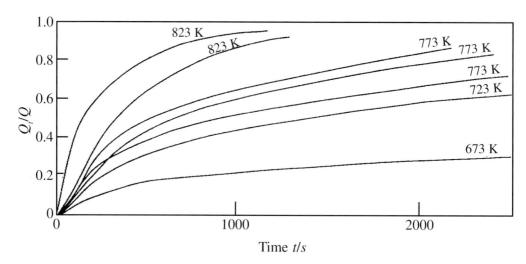

Fig. 7.5 Isothermal desorption of the prevailing hydrogen content from 12 mm diameter cylindrical samples machined coaxially from a commercially produced 16 mm diameter extruded rod of 99.99% pure aluminium. Diffusion proceeds in the presence of disseminated micropores constituting a field of diatomic hydrogen traps. Q_t/Q_∞ is the fraction desorbed after time, t.

Table 7.3 Effective Diffusion Coefficients, D', for Hydrogen in Manufactured Pure Aluminium Rod Derived by Matching Desorption Curves in Figure 7.5 to Equation 7.10 at Selected Successive Values of the Co-ordinates, $(Q_t/Q_\infty, t)$

Temperature T/K	Intrinsic Diffusion Coefficient[†] D/cm² s⁻¹	D'/cm² s⁻¹ Corresponding to Values of Q_t/Q_∞ Given Below				
		0.4	**0.5**	**0.6**	**0.7**	**0.8**
823	2.6×10^{-4}	1.8×10^{-4}	1.7×10^{-4}	1.7×10^{-4}	1.6×10^{-4}	1.6×10^{-4}
		1.3×10^{-4}	1.3×10^{-4}	1.2×10^{-4}	1.2×10^{-4}	1.2×10^{-4}
773	1.7×10^{-4}	5.4×10^{-5}	4.6×10^{-5}	4.4×10^{-5}	4.4×10^{-5}	4.7×10^{-5}
		3.4×10^{-5}	4.5×10^{-5}	3.8×10^{-5}	3.7×10^{-5}	3.8×10^{-5}
		2.7×10^{-5}	2.6×10^{-5}	2.5×10^{-5}	2.4×10^{-5}	2.5×10^{-5}
723	1.0×10^{-4}	1.9×10^{-5}	1.7×10^{-5}	1.7×10^{-5}	1.6×10^{-5}	1.7×10^{-5}
673	5.8×10^{-4}	3.6×10^{-6}	3.2×10^{-6}	3.0×10^{-6}	-	-

mistaken for the true diffusion coefficient, D. Using values of D' given by eqn 7.27 in place of D in solutions to Fick's equation describes the transport of hydrogen in manufactured products of moderate section reasonably well. It is also used as the basis for on-line data processing used to assist the vacuum extraction technique for hydrogen content determination described earlier in Section 5.2.4.

The effective diffusion coefficient, D' is always less than the corresponding true coefficient, D, and the difference becomes greater for progressively lower temperatures as illustrated by the comparison given in Figure 7.4, probably reflecting a temperature dependent change in the equilibrium between monoatomic hydrogen in solution and diatomic hydrogen traps as the solubility in the metal decreases. The much smaller deviations between replicate desorption experiments are probably not significant and reflect minor inconsistencies in empirical factors contributing to D'.

7.3.3 DIFFUSION WITH ENHANCED TRAPPING

The discussion in Section 7.3.2 applies to small samples or moderate sections of manufactured products in which removal of the hydrogen by desorption is virtually completed before changes in pore structure become significant. However, if the metal is subjected to prolonged heating with the existing hydrogen content retained in situ, pore structures can evolve according to the theoretical considerations in Section 4.2.1 with consequent redistribution of hydrogen between solution and pores. Such a situation can develop within thick sections of the metal if it is subjected to prolonged heat-treatments at high temperatures, which can last for up to 24 hours at 550°C in some production routes.

An experiment was devised[16] in which samples of pure aluminium were sealed against egress of hydrogen during prolonged heat-treatments to simulate the situation inside thick metal sections, e.g. thick plate or heavy extrusions. The seal was provided by a film produced on the metal surface by anodising in a non-solvent electrolyte. Samples, 150 mm long cut from a 19 mm diameter extruded rod of 99.99% pure aluminium were prepared to a fine surface finish and anodised in 0.2% sodium borate solution, progressively raising the potential to 500 V, yielding a film 700 nm thick.[18] Several of them were heated in a furnace with an air atmosphere at a controlled temperature of 550 ± 2°C and withdrawn in sequence after 4, 8, and 16 hours. The heat-treated samples were turned axially to provide cylinders 11 mm diameter × 110 mm long for hydrogen diffusivity measurements. These measurements were all conducted at a temperature of 773 K, using essentially the vacuum extraction method described in Section 7.2.2. The quantities of hydrogen collected showed that the hydrogen content which was 0.40 cm³/100 g was unaffected by the treatment.

The desorption curves, given in Figure 7.6, show that the evolution of hydrogen from metal that had experienced prolonged heat-treatment was initially slower than from metal in its original condition but it accelerated as desorption proceeded. The effect became progressively more pronounced as the time of prior heat-treatment was extended.

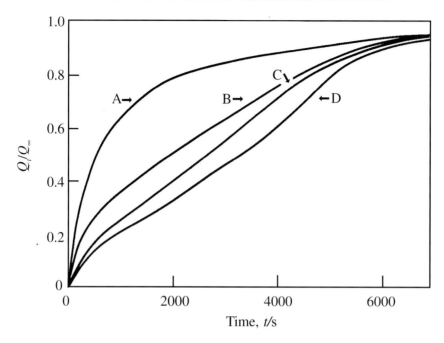

Fig. 7.6 Desorption at 773 K of the prevailing hydrogen content from cylindrical samples of a 99.99% pure aluminium extrusion. Effect of enhanced trapping induced by prolonged heat-treatment with the hydrogen sealed *in situ* by an anodic film. Q_t/Q_∞ is the fraction desorbed after time, t. Hydrogen content 0.40 cm³/100 g.

Curve A - Sample as extruded
          B - Sample preheated   4 hours at 823 K
          C - Sample preheated   8 hours at 823 K
          D - Sample preheated 16 hours at 823 K

Clearly, eqn 7.23 and related equations based on the concept of effective diffusion coefficients constants do not apply to the metal in the condition developed by prolonged heat treatment at high temperature with the hydrogen confined *in situ*. However, the pore stability diagram given in Figure 4.4 can offer some qualitative guidance on the observed sequence of events, as follows.

If the condition of the metal produced by heat treatment is represented by a point in the domain, AMB, in Figure 4.4, the initial desorption of hydrogen from the surface zones is offset by concurrent transfer of hydrogen in the rest of the metal from solution into the pores, because AMB is the domain of *pore growth*. This provisionally reduces the activity of hydrogen in solution driving diffusion, so that the solute mobility diminishes initially. With continuing loss of hydrogen, the locus of the point representing the system must ultimately reach the boundary of the domain AMC, in which hydrogen is returned to solution by spontaneous pore collapse so that its mobility recovers.

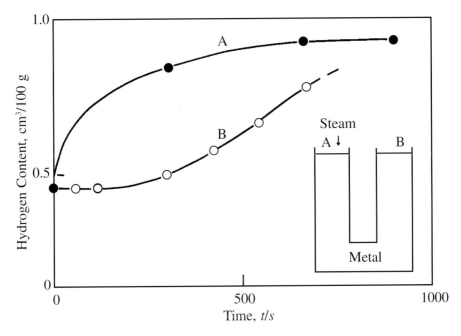

Fig. 7.7 Experiment to estimate the diffusivity of hydrogen in liquid pure aluminium. Hydrogen content at liquid metal surface in limb A of U-tube raised by steam injection. The plots show hydrogen contents of samples taken at intervals from metal surfaces in limbs A and B measured by vacuum extraction. (Ransley and Talbot).[10]

7.3.4 DIFFUSIVITY OF HYDROGEN IN PURE LIQUID ALUMINIUM

Reliable measurement of the diffusivity of hydrogen in liquid aluminium is difficult because it is subject to error from convection currents and interference from surface effects. Ransley and Talbot[10] determined an approximate value by the following method.

Liquid 99.8% pure aluminium was maintained at a uniform temperature of 740°C (1013 K) in an electrically heated refractory U-tube of 75 mm diameter, illustrated in Figure 7.7. The mean path through the metal between the surfaces, A and B in opposite sides of the tube was 700 mm. When the metal was in equilibrium with the ambient atmosphere, the hydrogen content at the surface of one side of the tube was raised by injecting steam. The hydrogen contents of samples taken at intervals from both surfaces were determined by vacuum extraction. Figure 7.7 gives plots of the hydrogen contents at both surfaces as a function of time. The diffusion coefficient estimated from these results is about 10^{-4} m^2 s^{-1} which is 3 or 4 orders of magnitude greater than the value for

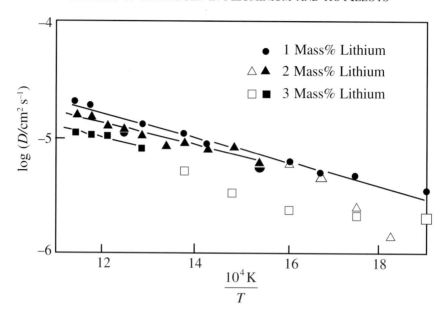

Fig. 7.8 Experimental values for diffusivities of hydrogen in binary aluminium-lithium alloys as functions of temperature (Anyalebechi *et al.*).[19] Solid points - α single phase field. Open points - (α + β) two-phase field.

the solid metal at its melting point, i.e. 5.8×10^{-4} cm^2 s^{-1}, obtained from eqn 7.25. The value is uncertain because error from convection currents could not be reliably assessed.

7.4 DIFFUSIVITY IN SOLID ALUMINIUM LITHIUM ALLOYS

7.4.1 Measured Values

Anyalebechi *et al.*[19] determined the diffusivities of hydrogen in binary alloys with 1, 2 and 3 mass% lithium by the isothermal absorption-quench-desorption method, as an integral part of the solubility measurements described earlier in Section 6.3.3.3, imposing the strict experimental disciplines mandated in Section 7.2.2. The materials were prepared from square rod hot-rolled by > 75% reduction from semi-continuously cast ingots. Cylinders 12 mm in diameter machined from the rod were consolidated by heat-treatment in vacuum for 9 hours at 873 K (600°C) and then re-machined dry to the final dimensions, 10 mm diameter × 50 cm long for the measurements.

The values obtained are given in Table 7.4 and plotted in Arrhenius format in Figure 7.8 Sufficient results are available for the Al - 1%Li and Al - 2%Li alloys at

Table 7.4 Temperature-Dependence of Hydrogen Diffusivity in Solid Binary
Aluminium-Lithium Alloys

Temperature T/K	Diffusion Coefficient[†], $D \times 10^6$/cm^{2}s^{-1}		
	Al – 1% Li	Al – 2% Li	Al – 3% Li
473	2.8	-	-
498	2.5	-	-
523	3.5	0.75	2.0
548	-	1.3	-
573	4.6	2.4	2.3
598	4.7	4.5	-
623	5.9	6.0	2.3
648	5.7	6.3	-
673	-	8.3	3.2
698	8.3	7.9	-
723	10.6	8.9	5.1
728	-	8.5	-
773	12.6	10.6	7.9
798	10.4	11.2	-
823	-	12.6	9.9
848	18.1	15.5	10.0
873	20.6	15.9	10.7

[†]Anyalebechi *et al.*[19]

temperatures above the solves temperatures to yield the following equations describing hydrogen diffusion in temperature ranges within which the alloys are single phase α:

Al - 1 mass% Li: 523 > T/K > 873: $D = 3.10 \times 10^{-8} \exp \{-2490/(T)\}$ cm^{2}s^{-1} (7.28)

Al - 2 mass% Li: 650 > T/K > 873: $D = 1.65 \times 10^{-8} \exp \{-2120/(T)\}$ cm^{2}s^{-1} (7.29)

Eqns 7.28 and 7.29 yield values of the activation energies for diffusion of 21 and 18 kJ mol^{-1} for the Al - 1 mass% Li and Al - 2 mass% Li alloys respectively, which are less than the value for true diffusion in pure aluminium,[8, 9] i.e. 45 kJ mol^{-1}. There are insufficient results in the temperature range between the solvus and solidus for similar analysis for the Al - 3 mass% Li alloy but the slopes of the plots in Figure 7.8 indicate that he activation energy is similar.

Table 7.5 Calculated and Observed Diffusion Coefficients for Hydrogen in Binary Aluminium-Lithium Alloys

| Alloy | T/K | Solubility cm³/100 g | | Partition Constant K | True Diffusivity‡ D/cm²s⁻¹ | Effective Diffusivity | |
		Ordinary Sites*	Preferred Sites†			Calculated D′/cm²s⁻¹	Observed§ D′/cm²s⁻¹
Al – 1% Li	873	0.026	0.85	32	3.8×10^{-4}	1.1×10^{-5}	1.8×10^{-5}
	823	0.015	0.77	51	2.6×10^{-4}	5.0×10^{-6}	1.5×10^{-5}
Al – 2% Li	873	0.026	0.98	37	3.8×10^{-4}	1.0×10^{-5}	1.5×10^{-5}
	823	0.015	0.88	59	2.6×10^{-4}	4.3×10^{-6}	1.3×10^{-5}

*Equation 6.7 (assumed as for pure aluminium)   †Equations 6.26 and 6.28
‡Equation 7.25 (assumed as for pure aluminium)   §Equations 7.28 and 7.29

The essential characteristics of the diffusion coefficients for hydrogen in the α phase are:
1. They are an order of magnitude lower than those for pure aluminium.
2. They are continuous across discontinuities in the solubility isobars in Figure 6.9.

7.4.2 MATHEMATICAL ANALYSIS

The partition of the solute between interstitial and cluster sites in aluminium-lithium alloys explained in Section 6.3.3.5 implies that the solute diffuses in a field of monoatomic traps. Therefore the measured values are probably best regarded not as true diffusion coefficients but as effective diffusion coefficients, D', as defined in Section 7.1.2.1:

$$D' = D/(K + 1) \tag{7.30}$$

where K is the quotient of equilibrium concentrations in interstitial sites and preferred sites.

In principle, values for D' can be calculated from eqn 7.30 for comparison with the measured values but in practice, neither D nor K can be determined explicitly.

Tentative estimates can be based on the following assumptions:
1. The true diffusivity, D in the Al-Li α phase is equal to that in the pure metal.
2. The solute is distributed between the preferred and ordinary sites is the ratio of the solubilities in the alloy, S(alloy) and in the pure metal, S(pure Al).

$$K = \frac{S(\text{alloy})}{S(\text{pure Al})}$$

Table 7.5 compares measured values with estimates made in this way for the Al - 1 mass% Li and Al - 2 mass% Li alloys at 873 and 823 K. There is fair agreement at

873 K but the calculated values seriously underestimate the effective diffusivity at 823 K. The discrepancy is most probably due to error in speculatively assuming that the solubility in ordinary sites of the α phase is equal to the solubility in pure aluminium.

7.5 REFERENCES

1. J. Crank, *The Mathematics of Diffusion*, Clarendon Press, Oxford, 1956.
2. R.M. Barrer, *Diffusion in and through Solids*, Cambridge University Press, Cambridge, 1941.
3. W. Jost, *Diffusion in Solids, Liquids, Gases*, Academic Press, London, 1952.
4. H.S. Carslaw and J.C. Jaeger, *Conduction of Heat in Solids*, Oxford University Press, Oxford, 1947.
5. A. Fick, *Ann. Phys. Lpz.*, **170**, 1855, 59.
6. J.B. Fourier, *Théorie Analytique de la Chaleur*, Oeuvres de Fourier, 1822.
7. P.N. Anyalebechi, *"The Solubility of Hydrogen in Pure Aluminium and Binary Aluminium-Lithium Alloys"*, Ph.D. Thesis, Brunel University, 1985.
8. W. Eichenauer and A. Pebler, *Z. Metallkünde*, **48**, 1957, 373.
9. W. Eichenauer, K. Hattenbach and A. Pebler, *Z. Metallkünde*, **52**, 1961, 82.
10. C.E. Ransley and D.E.J. Talbot, *Z. Metallkünde*, **46**, 1955, 328.
11. D.E.J. Talbot and D.A. Granger, *J. Inst. Metals*, **92**, 1963–4, 290.
12. C. Renon and J. Calvet, *Mem. Sci. Rev. Met.*, **58**, 1961, 835.
13. C. Renon and J. Calvet, *Mem. Sci. Rev. Met.*, **60**, 1963, 620.
14. Q.T. Fang and D.A. Granger, *Light Metals*, 1990.
15. Q.T. Fang, P.N. Anyalebechi and D.A. Granger, *Light Metals*, 1988, 477.
16. D.E.J. Talbot, *"Hydrogen in Aluminium and Aluminium Alloys"* Ph.D. Thesis, Brunel University, 1989.
17. D.E.J. Talbot, *International Met. Reviews*, **20**, 1975, 166.
18. M.S. Hunter and P. Fowle, *J. Electrochem. Soc.*, **101**, 1954, 481.
19. P.N. Anyalebechi, D.E.J. Talbot and D.A. Granger, *Met. Trans.*, **20B**, 1989, 523.

8. Absorption of Hydrogen in Aluminium and its Alloys

Hydrogen is produced by oxidation of aluminium in the presence of water vapour or hydrated materials with which it is in contact. The ultimate oxidation product is corundum, Al_2O_3, by the reaction:

$$2Al(\text{solid or liquid}) + 3H_2O(\text{gas}) = Al_2O_3(\text{solid}) + 3H_2(\text{gas}) \tag{8.1}$$

Eqn 8.1 formally describes the overall reaction but gives no information on its kinetics. The standard Gibbs free energy change for the reaction is:[1]

$$\Delta G^{\oplus} = -958387 - 72.87\ T \log T + 394.5\ T \text{ Joules (for solid aluminium)} \tag{8.2}$$

and $$\Delta G^{\oplus} = -979098 - 71.91\ T \log T + 413.6\ T \text{ Joules (for liquid aluminium)} \tag{8.3}$$

Applying the Van't Hoff isotherm at e.g. 700°C (973 K), assuming $a_{Al} = a_{Al_2O_3} = 1$:

$$\Delta G^{\oplus} = -RT \ln K = -RT \ln \frac{(p_{H_2})^3}{(p_{H_2O})^3} \tag{8.4}$$

$$\ln \frac{p_{H_2}}{p_{H_2O}} = -\frac{(-979098 - 71.91 \times 973 \log(973) + 413.6 \times 973)}{3 \times 8.314 \times 973} \tag{8.5}$$

whence: $$\ln \frac{p_{H_2}}{p_{H_2O}} = 1.2 \times 10^{14} \tag{8.6}$$

showing that virtually all traces of water contacting the metal are converted to hydrogen. It is impossible to eliminate water and its vapour completely from production environments and so aluminium products are vulnerable to hydrogen absorption during manufacture to an extent limited only by the opportunities afforded for reaction, the reaction kinetics discussed in Sections 8.3 and 8.4 and the ability of the metal to receive the gas.

Given this vulnerability, the design and operation of production processes are guided by principles that restrict and control hydrogen contents to acceptable values. The following brief summary anticipates the formal treatments in Chapters 9 and 10 and is given without explanation to provide a context for the current chapter.

Permissible hydrogen contents of liquid metal are not explicitly specified, but it is common practice to set local limits to avert macroscopic gas porosity in ingots or castings.

For most products a limit of $0.10 - 0.15$ cm^3/100 g suffices[2] but lower limits may be set on an application specific basis. If excessive, the hydrogen content is reduced before the liquid metal is cast by degassing it as described in Chapter 10 but there is a strong incentive to minimise the hydrogen content of the metal as presented to the degassing facilities to promote consistent and economic operation.

Solutions of hydrogen in solid aluminium and aluminium alloy products are invariably supersaturated because the solubility is so low and they can exert disruptive internal pressures. The threshold for hydrogen contents that can damage the metal depends on heat-treatments that the metal receives, which in turn depend on the differing requirements for various grades of pure metal and alloys and their applications. As a rough guide, solid metal with an initial hydrogen content of, say, 0.15 cm^3/100 g becomes more and more vulnerable as the hydrogen content is raised above about 0.3 cm^3/100 g during heat-treatment until as it approaches 1 cm^3/100 g the metal is damaged beyond recovery.

8.1 INDUSTRIAL HYDROGEN SOURCES

The predominant sources of hydrogen are water vapour in the atmospheres of furnaces used for melting or for heat treating solid products,[3-9] hydrated contaminants firmly bound to the surfaces of solid metal[3, 6, 7, 9–14] and imperfectly dried materials over which liquid metal passes during transfer from the furnace.

Atmospheric Water Vapour
Water vapour is usually present even in electric furnace atmospheres, from natural humidity and from lightly held water introduced on the surfaces of unprepared charge materials derived from lubricants, condensation and mildly hygroscopic detritus.

In furnaces fired directly with gas or oil, combustion products provide a copious supply of water vapour over the metal. The combustion products can be isolated in heat-treatment furnaces by indirect firing through burner tubes but with poor maintenance they can leak into the space containing the metal through cracked tubes and damaged seals.

Surface Contaminants on Solid Metal
Water that is chemically bound in contaminants firmly attached to the metal surface can persist to high temperatures. Corrosion and oxidation products usually contain hydrated aluminas typified by those listed in Table 8.1. On heating, they lose water as they pass through a series of structural changes, ultimately becoming the anhydrous oxide, corundum, but the transformation temperatures are so high that hydrated materials can serve as hydrogen sources at annealing, solution treatment and melting temperatures. Representative transformation temperatures at atmospheric pressure are:[15]

$$\text{Gibbsite} \xrightarrow{100°C} \text{Boehmite} \xrightarrow{500°C} \text{(Transition Aluminas)} \xrightarrow{1150°C} \text{Corundum} \quad (8.7)$$

Table 8.1 Representative Dehydration Sequence of Aluminas

Phase	Formula	Water Content (Mole Fraction)	Structure	Upper Transformation Temperature,°C
Gibbsite	$Al(OH)_3$	0.75	Monoclinic	100
Boehmite	$AlO(OH)$	0.5	Orthorhombic	500
η– Alumina	$Al_2O_3 x H_2O$	< 0.05*	Spinel	1150
Corundum	Al_2O_3	Anhydrous	Rhombohedral	–

*J. Anderson[21]

The most common contaminants are:
1. Resides from interaction between the metal and rolling mill lubricants that can contribute hydrogen not only to the solid metal during subsequent heat-treatments but also to liquid metal when some of it is remelted to recover the value of in-house rejections and trimmings.
2. Corrosion products on recycled scrap stored in unsuitable conditions.
3. Surface coatings, e.g. lacquer on recycled beverage cans.
   The degree of contamination relative to the mass of metal depends on its surface to volume ratio and is particularly significant for thin materials, e.g. rolled sheet. It is good practice to store light scrap under cover, clean it immediately before use, melt it by immersing it in liquid metal as compressed bales to minimise the period of short-range contact between the metal and the contamination.

Imperfectly Dried Materials Contacting Liquid Metal
Melting furnaces remain hot indefinitely during their campaigns, but the transfer systems conveying liquid metal from a furnace to moulds is dismantled after every batch is cast to allow the removal of ingots and reassembled before the next batch is cast. This includes refurbishment with wet refractory pastes that must be thoroughly dried out by gas burners. If the work is not properly carried out, the hot liquid metal subsequently flowing over the imperfectly dried material is exposed to steam from the residual water. Such carelessness is reprehensible, because it can inject hydrogen into the metal after it has been degassed.

8.2 THE CAPACITY OF THE METAL TO ACCEPT HYDROGEN

Liquid Metal
The capacity of the liquid metal to accept hydrogen is limited by its solubility at the prevailing atmospheric pressure and any excess is nucleated as gas bubbles that escape from the metal under their buoyancy. Some initial supersaturation is needed to nucleate

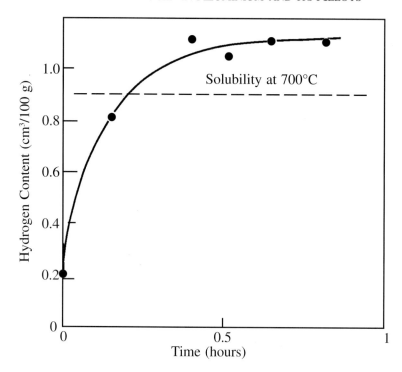

Fig. 8.1 Effect of steam injection on 99.8% pure aluminium at 700°C, showing that the hydrogen content rises to a limit only just above the solubility at the prevailing atmospheric pressure. (Ransley and Talbot).[3] (With permission, Gesellschaft für Metallkunde).

the gas but it is quite small. The results of Ransley and Talbot's experiments[3] given in Figure 8.1 show that even live steam injected into 25 kg of AA 1100 pure aluminium at 700°C failed to raise the hydrogen content above 1.1 cm³/100 g, i.e. just over the solubility, 0.88 cm³/100 g, where it was buffered by streams of bubbles carrying excess hydrogen out of the metal. In complementary passive experiments, small quiescent samples of the same metal held at 700°C could not be induced to accept more than 1 cm³/100 g of hydrogen when exposed to wet hydrogen at atmospheric pressure.

Solid Metal
In Section 4.2 of Chapter 4, the nucleation and growth of disseminated micropores was discussed in terms of the spontaneous breakdown of solutions but it has further significance in the context of hydrogen absorption, if the hydrogen content rises to the point at which the existing pore nuclei become critical. Transfer of hydrogen entering the metal into expanding pores then moderates the rise of hydrogen activity in solution, encouraging continuing absorption at the metal surface. At the same time, the pores provide extra capacity in the metal to accommodate the additional hydrogen. Hence the population of

potential micropore nuclei has a significant influence on the response of the metal to conditions in which hydrogen absorption is a possibility.

8.3 HYDROGEN ABSORPTION BY LIQUID METAL

8.3.1 PRACTICAL CONSIDERATIONS

8.3.1.1 Furnace Atmospheres

High hydrogen contents in liquid metal are naturally associated with high humidities and empirical practical observations bear this is out.[3–10] In former times, when techniques for removing hydrogen were inefficient, it was universally recognised that in temperate climates there was a seasonal factor in hydrogen-related damage to the metal; for example, surface blistering on flat-rolled products due to unsound ingots, was most prevalent in late Summer, when humidity is generally high and diminished sharply as Winter approached.

With an existing furnace design and prevailing climatic conditions, little can be done to control the water vapour pressure in the furnace atmosphere but its effect on the hydrogen content of the metal can be minimised by strict temperature control, because with rising metal temperature the solubility of hydrogen increases and reactions are accelerated. The effect is illustrated by Ransley and Talbot's observations[3] on 5 tonnes of AA 1100 pure aluminium heated slowly to 900°C in an oil fired furnace. The hydrogen contents determined by vacuum hot extraction of samples taken from the metal surface at temperature intervals of 50°C are given in Table 8.2. Within the accuracy of an industrial experiment, the ratios given in the fourth column show that the quantity of hydrogen in solution was proportional to the solubility.[16] Application of Sievert's isotherm, eqn 4.3, yields the notional activities of hydrogen gas and the corresponding pressures in equilibrium with the solutions given in the last two columns of the table. The hydrogen pressures, 3000 to 5000 Pa (0.03 to 0.05 atm.), were similar to the water vapour pressures measured in the furnace atmosphere. This unexpected apparent correlation is examined further in Section 8.3.2.5.

8.3.1.2 Charge Materials

Virgin Metal

To economise on fuel and time, there is an incentive to use virgin metal as liquid delivered from the smelter. The metal temperature on exit from the smelter is typically > 900°C, so that the hydrogen content is of the order of 0.6 cm^3/100 g due to exposure to the atmosphere at such extremely high temperature, as expected from Table 8.2. The metal is prepared for casting by adjusting its composition, removing gross detritus and reducing its temperature typically to 700–725°C to suit the casting arrangements. At the same time, the falling temperature provides an opportunity to exploit the diminishing solubility to

Table 8.2 Effect of Temperature on the Hydrogen Content in 5 tonnes of Liquid Pure Aluminium in an Oil-Fired Furnace[*]

Temperature, °C	Hydrogen Content (cm³/100 g)	Solubility[†] (cm³/100 g)	Ratio of Hydrogen Content to Solubility	Notional Hydrogen Activity	Equivalent Hydrogen Pressure p/Pa
750 (1023 K)	0.20	1.20	0.16	0.03	3000
800 (1073 K)	0.36	1.60	0.21	0.04	4000
850 (1123 K)	0.44	2.07	0.21	0.04	4000
900 (1173 K)	0.60	2.62	0.23	0.05	4000

[*]Ransley and Talbot[3] (with permission, Z.Metallkunde).

[†]Standard values for 101325 Pa assessed by Talbot and Anyalebechi.[16]

reduce the hydrogen content to a level within the capacity of subsequent degassing treatments. This is usually accomplished in two furnaces in tandem, one to receive the metal in batches from the smelter and the other to give time to condition it for casting.

Remelt Metal

Much aluminium and aluminium alloy production is from remelted metal, to provide stock for fabrication facilities remote from smelters, to recover the high value of in-plant recycled scrap or to produce special alloys for critical applications. When aluminium is remelted, the hydrogen content invariably rises.[2] There are so many variables in the origins and compositions of charge materials and the alloys produced from them that it is difficult to assemble, correlate and compare field observations from different plants. One common component of the charge, i.e. recycled rolling mill scrap has fairly consistent characteristics and is considered in Section 8.5 of this chapter.

8.3.2 REACTION KINETICS

8.3.2.1 Oxygen Activity Gradients in Oxidation Products on Aluminium

The hydrogen is absorbed during general oxidation of aluminium in humid atmospheres and the treatment of reactions producing it must be consistent with standard oxidation kinetics referred to the ionic character of metal oxides as close-packed assemblies of metal cations, M^{z+}, and oxygen ions, O^{2-}, where z is the oxidation state of the metal, M. If an oxide is hydrated, some or all of the oxygen ions may be protonated, becoming hydroxyl ions, OH^-.

If it is continuous and coherent, the oxidation product separates a metal from its environment but oxidation continues by the diffusion of ions derived from the primary reactants by partial reactions at the metal/oxide and oxide/atmosphere interfaces. The particular ions that diffuse depend on the pathways available. According to the defect structure of a particular oxide, the diffusing species may be (1) the metal, diffusing either as an interstitial ion, $M^{z+}\bullet$, or via vacant cation sites on the lattice, $M^{z+}\square$ or (2) oxygen diffusing via vacant anion sites on the lattice, $O^{2-}\square$; the charge is compensated by a simultaneous current of interstitial electrons, $e\bullet$, or electron holes, $e\square$, as appropriate to the diffusing species. If the oxidation product is permeated by hydroxyl ions, a special pathway is available for hydrogen as protons conducted by a Grotthus chain mechanism:

$$-OH^- -O^{2-} -O^{2-} - \quad \rightarrow \quad -O^{2-} -OH^- -O^{2-} - \quad \rightarrow \quad -O^{2-} -O^{2-} -OH^- - \quad \rightarrow \quad \text{etc.}$$

Wagner's classic theory applies to the usual situation in which rate-control is exercised by the diffusion so that local equilibrium is maintained at the oxide/atmosphere and metal/oxide interfaces.[17] For the oxidation of aluminium in air, the oxygen activities corresponding to the equilibria at the interfaces are:

Oxide/atmosphere interface
The activity equals that in the prevailing atmosphere, i.e. for normal air:

$$a_{O_2} = 0.21 \tag{8.8}$$

Metal/oxide interface
The activity is given by application of the Van't Hoff isotherm to the equilibrium between aluminium and corundum:

$$\frac{4}{3}Al + O_2 = \frac{2}{3}Al_2O_3 \tag{8.9}$$

for which:[1]

$$\Delta G^\ominus = -1132070 - 10.46\,T \log T + 257.3\,T \text{ Joules} \tag{8.10}$$

yielding the value:

$$a_{O_2} = 10^{-49} \text{ at } 700°C \text{ (973 K)} \tag{8.11}$$

These values span the range of oxygen activity that diminishes progressively through the thickness of the oxidation product from the oxide/atmosphere to the metal/oxide interface.

8.3.2.2 Conditions for Reduction of Water Vapour to Hydrogen
The Gibbs free energy for reduction of water vapour to hydrogen:

$$2H_2O = 2H_2 + O_2 \tag{8.12}$$

is given by:

$$\Delta G^\ominus = 492900 - 110\,T \text{ Joules} \tag{8.13}$$

In a humid atmosphere, e.g. with a water vapour partial pressure of 0.03 atm. corresponding to a dew point of 25°C, application of the Van't Hoff isotherm to eqn 8.13 at 700°C yields a critical value of oxygen activity, $a_{O_2} = 1.7 \times 10^{-24}$, below which hydrogen can be generated. Bearing in mind the limiting values of oxygen activity given in eqns 8.8 and 8.11, for the conditions considered, activities low enough to yield hydrogen are expected only in a narrow zone of the oxide immediately adjacent to the metal/oxide interface.

To proceed further, the structures and stabilities of phases that can form on aluminium oxidised in humid atmospheres must now be examined.

8.3.2.3 Relevant Phases in the Alumina-Water System

The alumina-water system includes the anhydrous oxide corundum α-Al_2O_3, trihydrates $Al(OH)_3$ e.g. gibbsite, monohydrates $AlO(OH)$ e.g. boehmite and various *transition aluminas*, that are intermediaries in the progressive dehydration of hydrates.[15] Transition aluminas are not true hydrates but have spinel or cubic structures stabilised by a small water fraction. In the present context, interest centres on corundum and a particular transition alumina, η-Al_2O_3, that can form on liquid metal.[15]

Corundum is a near-stoichiometric rhombohedral compound.[18] At temperatures < 1100 K it is a predominantly ionic conductor with a large gap between the valence and conduction bands, with some n and p-type electronic conduction.[19, 20] In particular, its structure is incompatible with diffusion of hydroxyl ions.

η-alumina is a 1:3 spinel, comprising an FCC lattice of O^{2-} ions with interstitial Al^{3+} and H^+ ions balancing the charge. A convenient crystallographic description of aluminium 1:3 spinels is based on a hypothetical prototype extended unit cell:

$$Al_8^{3+} \left[Al_{13\frac{1}{3}}^{3+} \square_{\frac{2}{3}} \right] O_{32}^{2-} \tag{8.14}$$

The cations, Al^{3+} before and inside the brackets occupy tetrahedral and octahedral interstitial sites respectively; the symbol, $\square$, represents the remaining vacant octahedral sites. Fractions in eqn 8.14 reconcile the stoichiometric ratios with the rectilinear unit cell. The hypothetical structure is unstable but it is stabilised by replacing some of the cations in octahedral sites with monovalent cations bearing an equivalent positive charge, e.g. replacing $1\frac{1}{3}$ Al^{3+} ion by four monovalent ions, X^+, yields:

$$Al_8^{3+} \left[Al_{12}^{3+} X_4^+ \right] O_{32}^{2-} \tag{8.15}$$

corresponding to the limiting composition for the phase, $5Al_2O_3 \cdot xH_2O$. An example is the sodium beta alumina, often written, $NaAl_5O_8$. In the special case of η-alumina with hydrogen as the monovalent species, a corresponding formula would be:

$$Al_8[Al_{12}H_4]O_{32} \tag{8.16}$$

but hydrogen cations are vanishingly small and can coordinate only with single oxygen ions, producing a structure permeated with hydroxyl groups, written schematically as:[21]

$$Al_8[Al_{12}](OH)_4O_{28} \tag{8.17}$$

8.3.2.4 Phase Equilibria

A definitive phase diagram for the alumina water system is elusive, due mainly to frustrating experimental difficulties. General trends can, however be indicated by a schematic diagram constructed from theoretical principles in the format of chemical potential diagrams that Ellingham devised for metal oxides and sulphides.[22]

Chemical Potential Diagrams

In these diagrams, standard Gibbs free energies of formation, $\Delta G^{\oplus}$, for related compounds formed from a common reactant, i, are plotted as functions of temperature, T. The purpose is to display their stabilities in sequence, by plotting the standard free energies *per mole of the common reactant*. If the compound and its precursor are pure, i.e. present at unit activity, equilibrium constants for the various reactions, K, depend only on the activity of the common reactant, a_i, and application of the Van't Hoff isotherm yields:

$$\Delta G^{\oplus} = - RT \ln K = - RT \ln a_i \tag{8.18}$$

The term, $- RT \ln a_i$ equals the chemical potential, μ_i of the common reactant.

If the common reactant is a gas, the temperature dependence of $\Delta G^{\oplus}$ can be estimated theoretically because it equals minus the standard entropy change, $-\Delta S^{\oplus}$, which is almost entirely due to the volume of gas consumed in the reaction,[22] i.e. 190 J $K^{-1}mol^{-1}$, so that $\Delta G^{\oplus}$ is a nearly linear function of temperature:

$$\frac{d\Delta G^{\oplus}}{dT} = - \Delta S^{\oplus} = -190\,\mathrm{JK}^{-1} \tag{8.19}$$

For the elevated temperatures and low pressures usually associated with chemical potential diagrams, the gas pressure is a reasonable approximation for the activity of the gas.

Application to the Alumina-Water System

Successive dehydration of species in the alumina water system can be represented in a water vapour potential diagram for the following reactions, written in the convention for chemical potential diagrams, i.e. for one mole of water with the gaseous species on the left:

$$AlO(OH)(s) + H_2O(g) = Al(OH)_3(s) \tag{8.20}$$

$$\frac{1}{1-N} (\eta - alumina)(s) + H_2O(g) = \frac{2}{1-N} AlO(OH)(s) \tag{8.21}$$

$$\frac{1}{N} Al_2O_3(s) + H_2O(g) = \frac{1}{N} (\eta - alumina)(s) \tag{8.22}$$

where N is the mole fraction of water in η-alumina.

According to eqn 8.19, plots of $\Delta G^{\oplus}$ against T representing equilibria for these reactions are all nearly straight lines with a common slope of 190 JK^{-1}. Their positions can be fixed by the transformation temperatures quoted in Table 8.1 that apply to the evolution of water vapour against atmospheric pressure, for which $\Delta G^{\oplus} = - RT \ln p_{H_2O} = 0$.

This information yields the schematic diagram given in Figure 8.2. It is correct in concept but not in detail because of assumptions made in its construction. Nevertheless, it is the only guide available with which to interpret the constitution of the oxidation products.

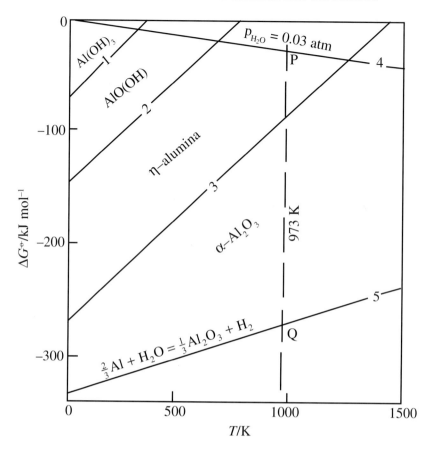

Fig. 8.2 Schematic water vapour potential diagram for some alumina hydrates based on estimated standard free energies of hydration and theoretical entropy changes.
Lines tracing equilibrium water vapour potentials for dehydration reactions:
1 - $Al(OH)_3 \rightarrow AlO(OH)$ 2 - $AlO(OH) \rightarrow \eta$-alumina   3 - η-alumina $\rightarrow Al_2O_3$
4 - Water vapour potential at the oxide/atmosphere interface, $p_{H_2O} = 0.03$ atm.
5 - Water vapour potential at the metal/oxide interface reaction imposed by the equilibrium
$\frac{2}{3}Al + H_2O = \frac{1}{3}Al_2O_3 + H_2$.

8.3.2.5 Correlation of Phase Relationships with Hydrogen Absorption

Theoretical Approach

The equilibrium lines divide the water vapour potential diagram into zones of stability for $Al(OH)_3$, $AlO(OH)$, η-alumina and α-Al_2O_3.

Further lines added to the diagram, trace out the water vapour potentials at the bounding interfaces of the oxide. The potential, μ_{H_2O}, at the oxide/atmosphere interface, is due to the prevailing water vapour pressure. The example plotted on the diagram is for $p_{H_2O} = 0.03$ atm. and the equation for the line is:

$$\mu_{H_2O}(\text{oxide/atmosphere interface}) = RT \ln (0.03) \tag{8.23}$$

and the potential at the metal/oxide interface is imposed by the equilibrium:

$$\tfrac{2}{3} Al + H_2O = \tfrac{1}{3} Al_2O_3 + H_2 \tag{8.24}$$

Since the activities of solid species are unity and for hydrogen notionally buffered by the external atmospheric pressure, $a_{Al} = a_{Al_2O_3} = a_{H_2} = 1$:

$$\mu_{H_2O}(\text{metal/oxide interface}) = \Delta G^{\oplus}(\text{per mol of water vapour})^1$$
$$= -326400 - 240 \, T \log T + 138 \, T \text{ Joules} \tag{8.25}$$

Consider the oxide forming on liquid metal at e.g. 700°C (973 K) marked by the vertical line on the diagram. The intercept, P, on the line representing the water vapour potential at the oxide/atmosphere interface is within the domain of stability for η-alumina. The corresponding intercept, Q, on the line representing the water vapour potential at the metal/oxide interface is within the domain of stability for corundum, α-Al_2O_3. Thus assuming local equilibrium at the oxide/air and metal/oxide interfaces, an oxide film growing in moist air is expected to be two-phase, with an outer layer of η-alumina and an inner layer of corundum, α-Al_2O_3.

Experimental Evidence

In Stephenson's experiments,[23] liquid AA 1070 (99.7% pure) aluminium was exposed to air with controlled humidity in the equipment illustrated in Figure 8.3. The hydrogen content of the metal was monitored continuously with a Telegas instrument for correlation with the constitution of oxide samples periodically lifted from the metal surface. To permit examination, metal adhering to the undersides of the samples was removed by selective dissolution in a solution of bromine in methanol. The initial hydrogen content was adjusted by purging the metal with steam to increase it or nitrogen to reduce it, as required. The metal surface was skimmed clean, and the prescribed atmosphere was established over it. Figures 8.4 to 8.8 record the hydrogen contents of the metal as functions of time of exposure for appropriate combinations of temperature and humidity; every figure gives a set of curves for different initial hydrogen contents within a range typical in industrial practice.

During the first 20 minutes of exposure to humid air, the metal rapidly absorbed hydrogen but this phase of the interaction was curtailed by the onset of a long term effect in which the hydrogen content approached a steady value characteristic of the metal temperature and prevailing humidity. If the initial upsurge in hydrogen content raised it above the ultimate steady value, it introduced a maximum in the hydrogen content, evident in the figures.

These effects correlate with the evolving structure of the oxidation product formed in humid air. According to Figure 8.2, the stable phase in contact with the metal is corundum but it does not form immediately and the initial oxidation product is η-alumina. Colonies of corundum subsequently grow from nuclei at the metal/oxide interface and

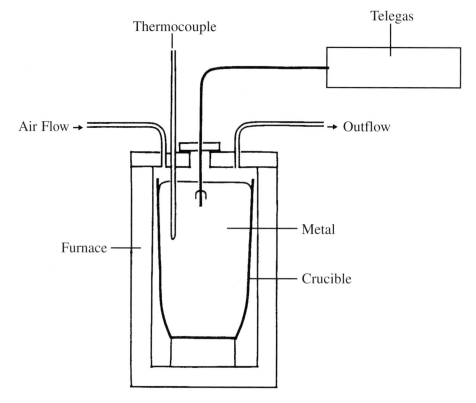

Fig. 8.3 Experimental equipment to monitor hydrogen absorption by liquid aluminium exposed to controlled humidities (Stephenson).[23]

after about 30 minutes at 700°C, they merge into a continuous layer over the metal surface underlying the initial layer of η-alumina. Corundum nuclei in a matrix of η-alumina are illustrated in the scanning electron micrograph in Figure 8.9 and the subsequent developing structure is illustrated by the transmission electron micrographs in Figure 8.10 given by Dholiwar.[24]

The direct contact between the metal and η-alumina during the delayed development of the corundum is responsible for the initial upsurge in hydrogen content, for the following reason. As Figure 8.2 indicates, η-alumina is unstable at the low water vapour potential near the metal/oxide interface and transforms to corundum. In the transformation, aluminium and oxygen ions are reordered and OH⁻ ions are eliminated, e.g. by the coupled reactions:

$$Al(\text{metal}) \rightarrow Al^{3+}(\text{oxide}) + 3e^-$$

$$3e^- + 3OH^-(\text{in } \eta\text{-alumina}) \rightarrow 3O^{2-} + 3H \quad\quad\quad (8.26)$$

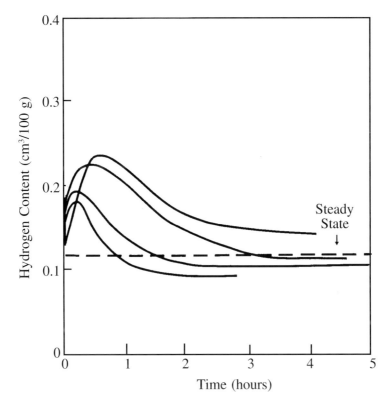

Fig. 8.4 Hydrogen content of liquid AA 1070 (99.7% aluminium) exposed to humid air. Temperature: 700°C. Water vapour partial pressure: 2000 Pa (0.02 atm.). The dotted line indicates the hydrogen content corresponding to a notional hydrogen activity equal to the activity of water vapour (Stephenson).[23]

The hydrogen is delivered close to the metal surface where it is available for dissolution. The supply is maintained by OH⁻ ions diffusing down the concentration gradient established by local depletion near the metal/oxide interface, perhaps from as far away as the oxide/atmosphere interface where they can be replenished from water vapour:

$$3H_2O(\text{atmosphere}) + 3O^{2-}(\text{oxide}) = 6OH^-(\text{oxide}) \tag{8.27}$$

When a complete cover of corundum is established over the metal/oxide interface, any interaction between the metal and hydroxyl ions must be transferred to the corundum/η-alumina interface, because corundum is impervious to hydroxyl ions. Conditions producing the steady state are difficult to analyse but probably include the following factors:

1. The oxygen potential at the corundum/η-alumina interface.

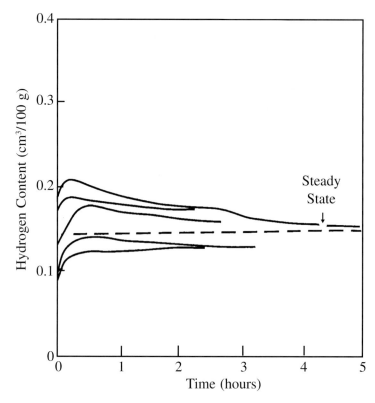

Fig. 8.5 Hydrogen content of liquid AA 1070 (99.7% aluminium) exposed to humid air. Temperature: 700°C. Water vapour partial pressure: 3000 Pa (0.03 atm.). The dotted line indicates the hydrogen content corresponding to a notional hydrogen activity equal to the activity of water vapour (Stephenson).[23]

2. Diffusion of aluminium ions, Al^{3+}, from the metal through the corundum layer.
3. Diffusion of hydrogen *atoms* to the metal through the corundum layer.
4. Diffusion of hydroxyl ions produced in Reaction 8.26 through the η-alumina.

In default of analysis, a pragmatic approach is possible, guided by the information from the industrial experiment described in Section 8.3.1.1, implying that in the long term, by whatever means it is established, the metal experiences a notional hydrogen activity equal to the activity of water vapour. Provisionally applying this idea to the conditions to which Figures 8.4 to 8.8 refer, Sieverts' isotherm yields the values for hydrogen contents given in Table 8.3 and indicated as the horizontal dotted lines superimposed on Figures 8.4 to 8.7. Their positions relative to the plots tracing the hydrogen contents correspond with the steady state values to which all of the hydrogen contents tend eventually. This apparently justifies the provisional assumption that in the long term, the notional hydrogen activity is controlled by the prevailing humidity. In dry

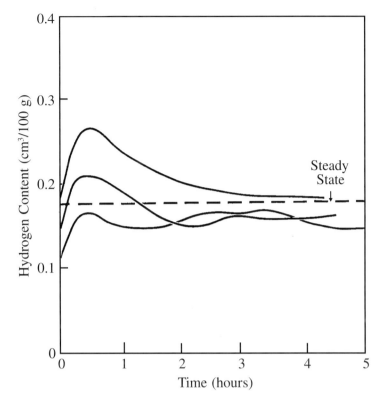

Fig. 8.6 Hydrogen content of liquid AA 1070 (99.7% aluminium) exposed to humid air. Temperature: 725°C. Water vapour partial pressure: 3000 Pa (0.03 atm). The dotted line indicates the hydrogen content corresponding to a notional hydrogen activity equal to the activity of water vapour (Stephenson).[23]

Table 8.3 Steady State Hydrogen Content of Liquid Aluminium in Humid Air Assuming a Notional Hydrogen Activity Equal to the Activity of Water Vapour

Temperature °C	Hydrogen Solubility* (cm³/100 g)	Water Vapour		Hydrogen Content (cm³/100 g)	Figure Reference
		Pressure p/Pa	Activity (aH$_2$O)		
700	0.88	2000	0.02	0.12	8.4
700	0.88	3000	0.03	0.15	8.5
725	1.03	3000	0.03	0.18	8.6
750	1.20	3000	0.03	0.21	8.7

*Assessed values for 101325 Pa (Talbot and Anyalebechi).[16]

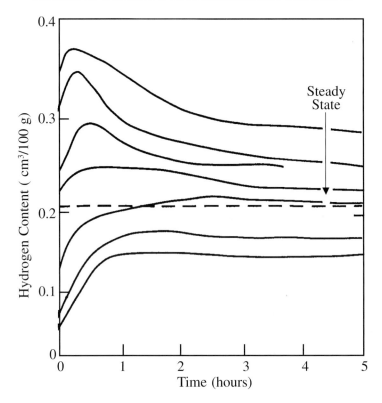

Fig. 8.7 Hydrogen content of liquid AA 1070 (99.7% aluminium) exposed to humid air. Temperature: 750°C. Water vapour partial pressure: 3000 Pa (0.03 atm.). The dotted line indicates the hydrogen content corresponding to a notional hydrogen activity equal to the activity of water vapour (Stephenson).[23]

air, η-alumina cannot form, the steady state is zero but it is approached slowly through the corundum oxidation product as illustrated by Figure 8.8.

8.4 ABSORPTION OF HYDROGEN IN SOLID METAL FROM THE ATMOSPHERE

8.4.1 PURE ALUMINIUM

In principle, solid pure aluminium with clean surfaces is susceptible to the absorption of hydrogen from atmospheric water vapour by the same process as the liquid metal, but in

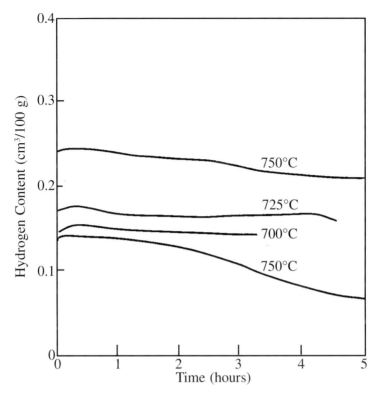

Fig. 8.8 Hydrogen content of liquid AA 1070 (99.7% aluminium) exposed to dry air (Stephenson).[23]

practice several factors intervene to restrict the uptake. It is impeded by the very low solubility of hydrogen in the solid metal and the much slower diffusion into the solid than into the liquid, with the result that the quantity absorbed is not significant by the time the process is terminated by the completion of the continuous layer of corundum at the metal surface.

Different criteria apply if the metal surface is contaminated, especially by subcutaneous detritus implanted in flat products by hot rolling, which provides a potent surface hydrogen source. It is an empirical extraneous effect unrelated to the chemical interaction between the metal and the atmosphere and is treated as a separate issue in Section 8.5.

Except for magnesium and lithium, the significant components of standard aluminium alloys form less stable oxides than the parent metal and do not contribute new phases to the oxidation products. Magnesium and lithium oxidise selectively, yielding new phases that can facilitate the absorption of hydrogen from atmospheric water vapour. Selective

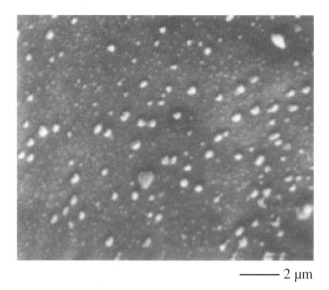

——— 2 μm

Fig. 8.9 Scanning electron micrograph of the underside of a film formed on liquid 99.99% pure aluminium after 10 minutes in air with 3 kPa (0.03 atm.) water vapour pressure, showing corundum crystals nucleating from η-alumina at the oxide/metal interface (Dholiwar).[24]

oxidation is approached by predicting the relative stabilities of the potential phases at the metal/oxide interface as functions of alloy composition. The stabilities of these phases and their influences are now considered.

8.4.2 ALUMINIUM-MAGNESIUM ALLOYS

8.4.2.1 Theoretical Treatment of Selective Oxidation

The oxide species to be considered are magnesium oxide (MgO), spinel ($MgAl_2O_4$) and corundum (Al_2O_3), all of which have been detected as oxidation products on aluminium-magnesium alloys.[25-28] Magnesium hydroxide need not be considered because it is unstable with respect to magnesium oxide for most conditions that the solid metal encounters in high temperature production processes, as illustrated in Table 8.4 for air with partial pressures of water vapour in the range 500 to 4000 Pa and temperatures in the range 300 to 600°C.

For any particular composition, the most stable oxide species at the metal/oxide interface is the one that can co exist with the metal at the lowest oxygen activity. This can be ascertained by comparing equilibrium oxygen activities for the oxides as functions of the magnesium content of the metal. The oxygen activities are calculated from the standard

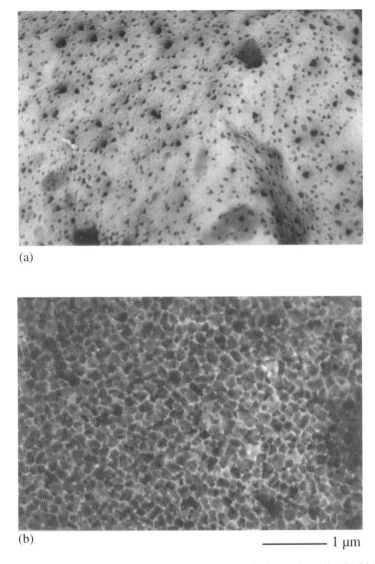

(a)

(b)   ——————— 1 μm

Fig. 8.10 Transmission electron micrographs of the oxide formed on liquid 99.99% pure aluminium in air with 3 kPa (0.03 atm.) water vapour pressure showing corundum, α-Al_2O_3 developing in a featureless η-alumina matrix. (a) after 1 minute and (b) after 30 minutes (Dholiwar).[24]

Gibbs free energies of formation of the oxides and the activities of magnesium and aluminium in the metal, as described in Appendix 2.

Figure 8.11 gives the results for a representative temperature, 500°C. The plots for $[a_{O_2}]_{Al_2O_3}$, $[a_{O_2}]_{MgAl_2O_4}$ and $[a_{O_2}]_{MgO}$ intersect at two critical compositions, 0.16 and 1.05%

Table 8.4 Gibbs Free Energy Changes for the Reactions of Magnesium and Lithium Oxides with Atmospheric Water Vapour[*]

Reaction	Water Vapour		Temperature		$\Delta G/J^{\dagger}$	Stable Species
	p/Pa	a_{H_2O}	°C	T/K		
MgO + H$_2$O = Mg(OH)$_2$	4000	0.04	300	573	+ 31078	Oxide
	4000	0.04	400	673	+ 44534	Oxide
	4000	0.04	500	773	+ 57990	Oxide
	4000	0.04	600	873	+ 71447	Oxide
	500	0.005	300	573	+ 36522	Oxide
	500	0.005	400	673	+ 50928	Oxide
	500	0.005	500	773	+ 65334	Oxide
	500	0.005	600	873	+ 79740	Oxide
Li$_2$O + H$_2$O = 2LiOH	4000	0.04	300	573	− 69086	Hydroxide
	4000	0.04	400	673	− 55630	Hydroxide
	4000	0.04	500	773	− 42174	Hydroxide
	4000	0.04	600	873	− 28718	Hydroxide
	500	0.005	300	573	− 63642	Hydroxide
	500	0.005	400	673	− 49236	Hydroxide
	500	0.005	500	773	− 34830	Hydroxide
	500	0.005	600	873	− 20425	Hydroxide

[*]Using eqns A2.21 and 2.22 in Appendix 2. [†]Per mole of water vapour.

magnesium, defining ranges in which each of the oxides establishes the lowest oxygen potential i.e.:

Magnesium content, %	0 – 0.16	0.16 – 1.05	> 1.05
Oxide co-existing at the lowest oxygen activity	Al$_2$O$_3$	MgAl$_2$O$_4$	MgO

To check whether the calculations correctly indicate the oxide which actually forms, Silva and Talbot[29, 30] conducted a systematic series of experiments in which small samples of binary aluminium-magnesium alloys with magnesium contents in the range, 1 to 9 wt.% were oxidised at 500°C for periods of up to 2 hours, taking care to ensure that the oxidation was truly isothermal and that the observations were not influenced by pre-existing films on the sample surfaces. Thorough examination of oxides detached from the samples by X-ray diffraction, electron diffraction and elemental analysis by X-ray photon spectroscopy invariably revealed magnesia but failed to reveal spinel or alumina on any sample. This is consistent with the prediction that magnesia forms at the metal surface

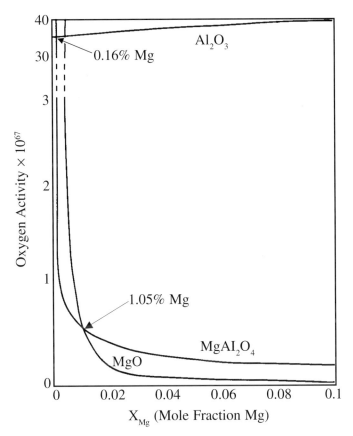

Fig. 8.11 Oxygen activity in equilibrium with Al_2O_3, $MgAl_2O_4$ and MgO on aluminium-magnesium alloys at 500°C as functions of magnesium content. Note change of scale for oxygen activity in equilibrium with Al_2O_3. Critical compositions expressed as magnesium %.

for alloys with magnesium contents > ~ 1%. Magnesia may persist as a metastable product at somewhat lower magnesium contents because the complex spinel structure, $MgAl_2O_4$ is probably more difficult to nucleate at the metal surface. Changes in the critical compositions in the temperature range 350 to 550°C are not large, supporting the view that magnesia is the characteristic high temperature oxide for binary alloys with magnesium contents ≥ 1 wt.%.

8.4.2.2 Hydrogen Absorption in Aluminium Magnesium Alloys from Clean Humid Air

The hydrogen potential applied to the metal/oxide interface through the medium of the particular oxidation product on the metal surface is obviously important in determining the absorption of hydrogen but it is not the only factor; the capacity of the metal structure to accommodate it is also important.

The Capacity of the Metal for Hydrogen

It has long been recognised that wrought aluminium and aluminium alloy products almost invariably contain microvoids, typically 1 – 2 μm in diameter, widely disseminated throughout the structure. Renon and Calvet[31, 32] first noticed them and Talbot and Granger[33] subsequently identified them as hydrogen filled micropores, generated during the solidification of their ingot precursors and persisting through fabrication. They appear typically as illustrated in Figures 9.11 to 9.14 and 9.16 to 9.18, given in Chapter 9. The microvoids can make a vital contribution to the capacity of the metal for hydrogen, because they offer potential nucleation sites for an internal gas phase to supplement the hydrogen in solution.

Quantities of Hydrogen Absorbed from a Humid Atmosphere

Morton[34] assessed the roles of the magnesia film and of potential pore nuclei in determining the quantity of hydrogen absorbed from humid air. The alloy used was AA 5052 which has a magnesium content well above the critical value required to form magnesia as the oxide. It was a standard commercial product in the form of 13 mm diameter extruded rod with the analysis, 2.20% Mg, 0.25% Si, 0.34% Fe, 0.10% Cu and an initial hydrogen content of 0.16 cm³/100 g.

As expected, the material contained an abundance of potential pore nuclei in the form of microvoids. To assess the role of the pore nuclei, standard samples of the same metal without them were required for comparison. These samples were prepared by extracting the initial hydrogen content at 500°C in vacuum as described in Section 5.2, so that the nuclei were unsupported and collapsed to annihilation under surface tension. During the extraction of hydrogen, a thin subcutaneous zone was depleted of magnesium by evaporation and it was removed by machining to restore the original surface composition. For convenience, samples with and without pore nuclei are distinguished by the terms unconsolidated and consolidated samples respectively. Machined cylindrical samples, 10 mm diameter × 60 mm of both kinds were prepared.

Identical experiments were conducted with unconsolidated and consolidated samples. They were heated at prescribed temperatures in flowing clean air with a water vapour partial pressure of 3050 Pa (0.03 atm.) introduced by saturating it at the corresponding dew point, 25°C. Samples were withdrawn successively at intervals of 1 hour over a period of 6 hours, which is sufficient to relate to industrial heat-treatments. Experiments were conducted at 500 ± 2 and 550 ± 2°C. The hydrogen contents of the treated samples were determined by vacuum extraction, removing the minimum of material in sample preparation compatible with reliable determinations. In this way, surface enrichment of the hydrogen content was included in the results. The results are plotted in Figures 8.12 and 8.13, which exhibit the following features:

1. Curves tracing the hydrogen content of consolidated samples rise progressively but only to low values, ~ 0.05 cm³/100 g at 500°C and ~0.07 cm³/100 g at 550°C in 6 hours.

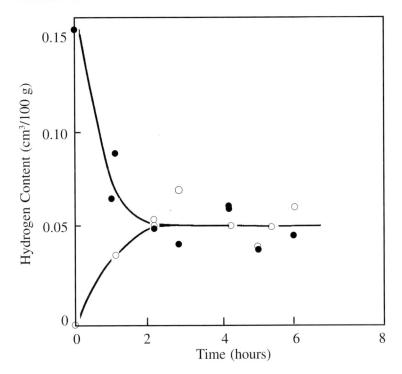

Fig. 8.12 Influence of clean humid air on the hydrogen content of an aluminium-magnesium alloy at 500°C.

Metal: 10 mm diameter samples machined from an extruded rod of an aluminium-2.2 wt.% magnesium alloy.

Atmosphere: Clean air with water vapour at a partial pressure of 3050 Pa (0.03 atm.).

●   Unconsolidated samples in their original manufactured state.
○   Consolidated samples with initial hydrogen content extracted.

2. Corresponding curves for unconsolidated samples first *fall* to meet the corresponding curves for consolidated samples and thereafter coincide with them.

The limited response to atmospheric humidity is perhaps inconsistent with a common perception that aluminium-magnesium alloys are prone to absorb hydrogen rapidly but nevertheless it is consistent with well-documented characteristics of *pure* magnesia,[35-40] expected as the stable oxide. The volume ratio for magnesia on aluminium alloys is close to unity, and it is formed as a coherent film on the metal. Moreover it is an ionic conductor, in which intrinsic diffusion is very slow[38] so that the pure oxide is expected to inhibit oxidation and related processes. Further thought is needed to identify a mechanism whereby hydrogen can be transmitted to the metal through the oxidation product on its surface.

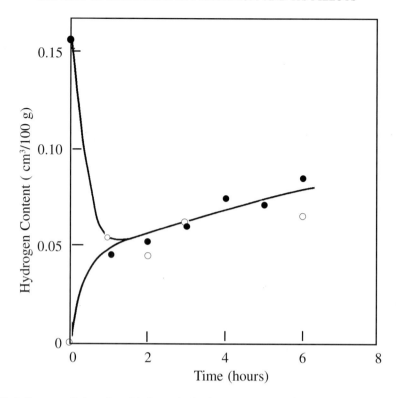

Fig. 8.13 Influence of clean humid air on the hydrogen content of an aluminium-magnesium alloy at 550°C.

Metal: 10 mm diameter samples machined from an extruded rod of an aluminium-
                 2.2 wt.% magnesium alloy. (as in Figure 8.12).

Atmosphere: Clean air with water vapour at a partial pressure of 3050 Pa (0.03 atm.). (as in
                 Figure 8.12).

●   Unconsolidated samples in their original manufactured state.
○   Consolidated samples with initial hydrogen content extracted.

Reference to the information in Table 8.4 shows that magnesium hydroxide is highly unstable with respect to magnesia in the presence of the water vapour content of humid air and therefore cannot replace it as a bulk phase by the hydration reaction:

$$MgO(s) + H_2O(g) = Mg(OH)_2(s) \tag{8.28}$$

for which the standard Gibbs free energy is:[1]

$$\Delta G^{\ominus} = -46024 + 100.1T \text{ Joules} \tag{8.29}$$

but there is an alternative interaction between MgO and water vapour. It is well known that hydroxyl ions derived from atmospheric water vapour can enter the MgO structure,[40, 41] by reacting with to surface oxygen ions to form hydroxyl ions that can migrate inwards by a proton chain mechanism:

$$H_2O + O^{2-} = 2OH^-$$

$$OH^- + O^{2-} + O^{2-} \rightarrow O^{2-} + OH^- + O^{2-} \rightarrow O^{2-} + O^{2-} + OH^- \rightarrow \quad \text{etc.} \quad (8.30)$$

The surface hydroxyl ion population cannot be calculated rigorously because energy terms arising from the disturbance to the MgO simple cubic lattice are unknown but eqn 8.29 can be used as a guide to estimate the order of magnitude. Applying the Van't Hoff isobar to the expression for $\Delta G^{\ominus}$ in the equation yields the value, $K = 0.0055$ for the equilibrium constant at 500°C. Assuming $a_{H_2O} = 0.03$, $a_{MgO} = 1$ and $T = 500°C$, as in Morton's experiments, indicates that the OH$^-$ ion activity is of the order 10^{-4}. In principle, this provides a path for OH$^-$ ions but because their concentration is so low, their arrival and reduction to hydrogen at the metal/oxide interface is expected to be tangible but slow.

An explanation for the contrast between the responses of consolidated and unconsolidated samples is deferred to Section 8.4.2.4, where it is considered together with further information clarifying the role of potential pore nuclei, given in Section 8.4.2.3.

8.4.2.3 Hydrogen Absorption from Air Polluted with Sulphur Dioxide

There is a well known industrial problem that is sometimes encountered when medium strength age-hardening aluminium–magnesium–silicon alloys are solution-treated in air furnaces. It is manifest as severe hydrogen blistering with thick discoloured films on the metal surface and it is associated with an odour of hydrogen sulphide emitted by any metal that escapes quenching and cools in air of normal humidity. The conclusion to be drawn from these observations is that the effect is due to a change in the character of the oxidation product in the presence of sulphur which stimulates the absorption of hydrogen. There are several sources of sulphur contamination in furnaces, e.g. leaked products of combustion and the decomposition of fabrication lubricants carried in on the metal. In the oxidising conditions inside an air furnace, any contaminating species is converted to sulphur dioxide.

Hydrogen Absorbed in the Presence of Sulphur Dioxide

To quantify the effect, Morton[34] introduced sulphur dioxide at a partial pressure of 1000 Pa (0.01 atm) into the air flow in further experiments that were otherwise identical to those recorded in Figures 8.12 and 8.13, yielding the results given in Figures 8.14 and 8.15. Much greater quantities of hydrogen were absorbed than for corresponding thermal treatments without sulphur dioxide and comparison with Figures 8.12 and 8.13 demonstrates the scale of the effect. The most influential factor was the initial condition of the metal. Much more hydrogen was absorbed by unconsolidated than by consolidated samples.

The samples increased in diameter and were covered with surface blisters to an extent determined by the quantity of hydrogen absorbed. The cause was the expansion of internal spherical pores disseminated throughout the metal, revealed in electropolished

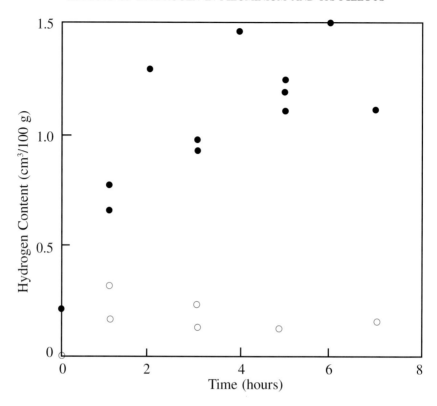

Fig. 8.14 Influence of sulphur dioxide contamination on the hydrogen content of an aluminium-magnesium alloy exposed to humid air at 500°C.

Metal: 10 mm diameter samples machined from an extruded rod of an aluminium-2.2 wt.% magnesium alloy. (as in Figures 8.12 and 8.13).

Atmosphere: Air with partial pressures of 3050 Pa (0.03 atm.) of water vapour and 1000 Pa (0.01 atm.) of sulphur dioxide.

●   Unconsolidated samples in their original manufactured state.

○   Consolidated samples with initial hydrogen content extracted.

microsections, as in the example given in Figure 8.16. Table 8.5 records the developing volume of porosity calculated from the densities of the samples measured by Archimedes method referred to the density of a sample consolidated by an equivalent thermal cycle in vacuum. Although convenient for resolving the porosity clearly, electropolishing exaggerates the quantity present as determined by quantitative image analyses[42] of microsections also reported in the table.

Characteristics of the Surface Films

The introduction of sulphur dioxide into the atmosphere transformed the character of the oxidised surface. The metal surface developed blisters and a grey patina, illustrated in

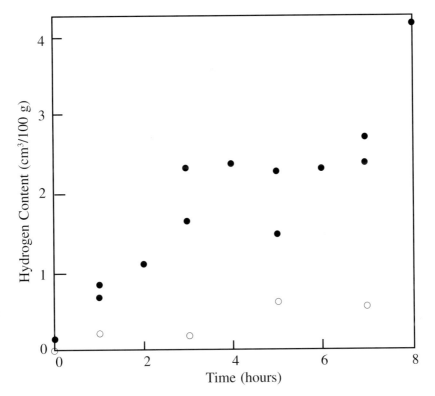

Fig. 8.15 Influence of sulphur dioxide contamination on the hydrogen content of an aluminium-magnesium alloy exposed to humid air at 550°C.

Metal: 10 mm diameter samples machined from an extruded rod of an aluminium-2.2 wt.% alloy. (as in Figures 8.12 and 8.13).

Atmosphere: Air with partial pressures of 3050 Pa (0.03 atm.) of water vapour and 1000 Pa (0.01 atm.) of sulphur dioxide.

●   Unconsolidated samples in their original manufactured state.

○   Consolidated samples with initial hydrogen content extracted.

Figure 8.17, thick enough to exhibit interference colours that darkened progressively, eventually becoming nearly black. While hot, it was odourless but when it was allowed to cool in air, there was an odour of hydrogen sulphide that persisted for several days. Under the microscope, small, isolated faceted crystals were visible on the surface of the film.

Assessment of the Oxidation Product

Consider how the oxidation product might be influenced by introducing sulphur dioxide. The first step is to assess the notional replacement reaction:

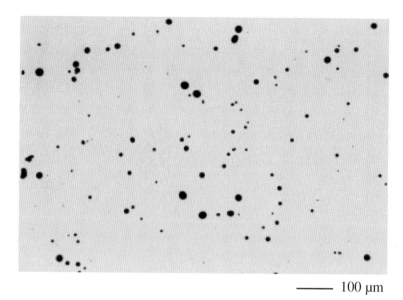

——— 100 µm

Fig. 8.16 Spherical porosity generated throughout a Al - 2 mass% Mg alloy rolled rod 10 mm in diameter, heated for 7 hours at 500°C in air with partial pressures of 3050 Pa (0.03 atm) water vapour and 1000 Pa sulphur dioxide. Electropolished section.

Table 8.5 Growth of Porosity in Aluminium-2 wt.% Magnesium Alloy Heated at 500°C in air with 3050 Pa (0.03 atm) of Water Vapour and 1000 Pa (0.01 atm) of Sulphur Dioxide

Time at Temperature t (hour)	% Porosity		
	Density Measurements		Quantitative Image Analysis
	Replicates	Average	
0	0.007, 0.009, 0.027	0,012	0,124
1	0.079, 0.071, 0.101	0.084	0.365
2	0.179, 0.172, 0.198	0.183	0.438
3	0.155, 0.169, 0.201	0.173	0.550
4	0.202, 0.180, 0.225	0.202	0.947
5	0.243, 0.241, 0.290	0.258	0.707
6	0.228, 0.243, 0.259	0.243	0.957
7	0.293, 0.315, 0.324	0.311	1.150
8	0.243, 0.224, 0.266	0.244	1.485

Fig. 8.17 Patina and blisters formed on 6 mm diameter rod of AA 6063 alloy (Al - 0.65% Mg- 0.50% Si alloy) heated 4 hours at 540°C in air with partial pressures of 3050 Pa water vapour and 1000 Pa sulphur dioxide.

$$MgO(s) + SO_2(g) = MgS(s) + \tfrac{3}{2}O_2(g) \qquad\qquad (8.31)$$

The Gibbs free energy change for the reaction is derived from the Gibbs free energies of formation[1] for MgS, MgO and SO$_2$, i.e.:

$$\Delta G^{\ominus} = 549442 + 12.3T \log T - 119\,T \text{ Joules} \qquad\qquad (8.32)$$

Substituting for T in eqn 8.32 yields large positive values for $\Delta G_1^{\ominus}$ at all reasonable temperatures, showing that MgS is so unstable with respect to MgO in the presence of sulphur dioxide that its formation in preference to the oxide can be discounted. Aluminium sulphide, Al$_2$S$_3$ is even less stable than MgS, so that there is no question of replacing the oxide film with a sulphide.

There is, however, an alternative interaction. It is well known that sulphur dioxide contamination accelerates the oxidation of metals in air,[43] sometimes by orders of magnitude by replacing a minority of the O^{2-} ions in oxide structures with S^{2-} ions. The replacement of O^{2-} ions in MgO with S^{2-} ions is expected to diminish its ionic nature, enhancing its extrinsic character, thereby facilitating both the self diffusion of magnesium and the diffusion of foreign species, including OH^- ions. This can form the basis of explanations for both an accelerated rate of oxidation and accelerated transport of OH^- ions that is a prerequisite for hydrogen absorption.

Such a kinetic view is consistent with the characteristics of the oxidation product stimulated by sulphur dioxide, e.g. the presence of S^{2-} ions in the oxide is manifest as the evolution of hydrogen sulphide by the hydrolysis:

$$S^{2-} + H_2O = O^{2-} + H_2S \qquad\qquad (8.33)$$

and the interference colours observed show that the oxide film thickens rapidly.

8.4.2.4 The Influence of Pore Nuclei

The information in Figures 8.12 to 8.15 can be correlated and explained using the logic developed in Chapter 4, Section 4.2.

In any particular thermal treatment, the quantity of hydrogen absorbed or retained in the metal is determined by the combination of two parameters:

1. The hydrogen activity produced by the environment at the metal/oxide interface.
2. The capacity of the metal to receive the gas.

The capacity for hydrogen of consolidated metal, assumed to be perfectly sound, is confined to solution in the metal, because there is nowhere else for it to go. The same metal in its unconsolidated state has a potential additional capacity for occluded hydrogen by virtue of its ability to accommodate it in pores as an internal gas phase, if the hydrogen activity is sufficient to expand the multiple pore nuclei that it contains. The conditions that determine whether or not this latent accommodation for hydrogen is activated can be described using Figure 4.4 in Chapter 4.

Three situations are distinguished:

1. If no pore nuclei are present, Figure 4.4 does not apply and the capacity of the metal for hydrogen is limited to its solubility in equilibrium with the prevailing hydrogen activity, which is generally low for aluminium alloys.* This situation fits the response of consolidated metal denoted by the open symbols in Figures 8.12 to 8.15, where the hydrogen content slowly approaches saturation values. More hydrogen is absorbed for the higher hydrogen potential imposed by sulphur laden humid air and more at 550°C than at 500°C, because the solubility rises with temperature and hydrogen activity in accordance with eqn 6.1 in Chapter 6.
2. If the coordinates representing hydrogen content and pore radius remain within the domain *below AMC* any vestige of an internal gas phase is unstable. This explains the loss of hydrogen from unconsolidated metal at the low hydrogen potential imposed by clean humid air because the pore nuclei collapse expelling the hydrogen they contain.
3. If the co ordinates enter the domain *above AMB*, the pore nuclei expand to a stable radius at the appropriate point on MB, corresponding to the minimum in the Helmholtz function in Figure 4.3b. The capacity of the metal is thereby increased by the quantity of hydrogen needed to occupy the expanded pore system. The response of unconsolidated metal denoted by solid circles in Figures 8.14 and 8.15 is consistent with the activation of this enhanced capacity, both in the quantities of hydrogen absorbed and in the observed development of internal porosity.

---

*Except for the special case of alloys containing lithium.

On first consideration, these arguments may seem academic and convoluted but they have the following very important practical implications for metal quality:

1. Damage to the metal by porosity generated during thermal treatment is not a consequence of hydrogen absorption but an integral part of the absorption process itself.
2. Pore nuclei in wrought metal are residues of unsoundness inherited from earlier operations in the production route, notably in casting as explained in Chapter 9. Therefore improving the quality of intermediate forms in early stages of production reduces the vulnerability of the metal in adverse environments in later thermal treatments.

8.4.2.5 Inhibition by Potassium Borofluoride

It is well-known from Stroup's patents,[44] that blistering experienced on aluminium-magnesium alloys solution-treated in industrial forced air circulation furnaces can be ameliorated by placing a small quantity of a slightly volatile fluoride, usually potassium tetrafluoroborate, KBF_4, in a hot zone of the furnace.[45, 46]

The function of potassium tetrafluoroborate is to suppress hydrogen absorption in the presence of sulphur dioxide, as the following experiments[46] demonstrate for AA 6063, an alloy typical of a range of medium strength alloys often solution-treated as long tubes and sections in forced air circulation furnaces; the material was in the form of 6.4 mm diameter rod with the analysis 0.65% Mg, 0.50% Si. Samples 6.0 mm in diameter × 40 mm long were machined to remove the rolled surface. They were solution-treated at the recommended temperature, 540°C for periods of up to 4 h in each of the following environments:

1. Flowing humid air with a water vapour partial pressure of 3050 Pa (0.03 atm.).
2. As for 1 but with sulphur dioxide injected at a partial pressure of 1000 Pa (0.01 atm.).
3. As for 2 but in the presence of the vapour from a small block of fused potassium tetrafluoroborate placed close to the hot samples.

The hydrogen contents were measured by vacuum extraction. Figure 8.18 compares the responses of the metal to the three different environments. The presence of the hot potassium tetrafluoroborate was very effective in suppressing hydrogen absorption in the presence of sulphur dioxide. The samples so treated were perfectly sound and instead of the heavy sulphur laden oxidation product otherwise expected, the metal surface had a light silvery sheen. The implication is that the protective mechanism is to produce a surface condition that is impervious to OH^- ions. The active agent is boron trifluoride gas evolved from potassium tetrafluoride, which decomposes at temperatures >350°C under atmospheric pressure:[47]

$$KBF_4 = KF + BF_3 \qquad (8.34)$$

Boron trifluoride is one of the strongest Lewis acids (electron acceptors) known[48] and can bind to electron donors, such as unsaturated O^{2-}, OH^- and S^{2-} ions exposed at the surface of an impure oxide. Thus the film is probably sealed by a boron/fluoride based surface condition but it is too thin to identify positively.

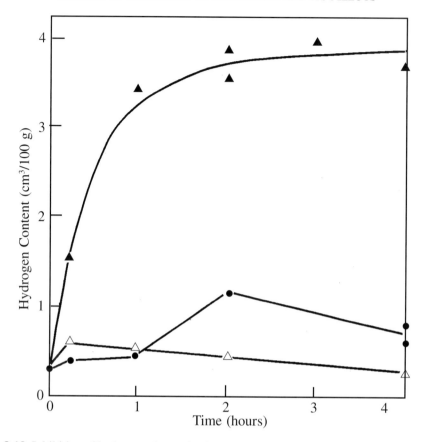

Fig. 8.18 Inhibition of hydrogen absorption by potassium tetrafluoroborate during solution-treatment of an age-hardening aluminium-magnesium-silicon alloy.

Metal: 6 mm diameter unconsolidated samples machined from AA 6063 alloy (Al - 0.65% Mg - 0.50% Si) rolled rod.

Atmosphere:

●   Air + 3050 Pa (0.03 atm.) partial pressure of water vapour.

▲   Air + 3050 Pa (0.03 atm.) partial pressure of water vapour + 1000 Pa (0.01atm.) partial pressure of sulphur dioxide.

△   Air + 3050 Pa (0.03 atm.) partial pressure of water vapour + 1000 Pa (0.01 atm.) partial pressure of sulphur dioxide + vapour from potassium tetrafluoroborate at 540°C.

8.4.3 ALUMINIUM-LITHIUM ALLOYS

8.4.3.1 *Theoretical Treatment of Selective Oxidation*

The possible surface oxidation products to be considered for the aluminium-lithium-oxygen-water system are corundum (Al_2O_3), lithium oxide (Li_2O), the 1:3 spinel ($Li_2Al_2O_4$) and lithium hydroxide. The relative stability of these species as functions of

Table 8.6 Standard Gibbs Free Energies of Formation per Mole of Oxygen for Aluminium, Magnesium and Lithium Oxides, Spinels and Hydroxides[*]

Temperature		$\Delta G^{\ominus}$/J mol^{-1} (O$_2$)						
°C	T/K	Al$_2$O$_3$	MgO	MgAl$_2$O$_4$	Mg (OH)$_2$	Li$_2$O	Li$_2$Al$_2$O$_4$	LiOH
400	673	–974571	–1063800	–1010023	–1023247	–1021464	–1045965	–1167229
500	773	–953776	–1043533	–989468	–980981	–993594	–1024251	–1117574
600	873	–933050	–1023406	–968985	–940854	–965242	–1002464	–1067437

[*]Calculated using eqns A2.1 to A2.11 in Appendix 2.

alloy composition cannot be determined explicitly as they can for the aluminium-magnesium system because values are not available for lithium and aluminium activities in the α solid solution. However a comparison of their standard Gibbs free energies of formation gives some guidance on which of them is expected as surface oxidation products. These quantities together with corresponding quantities for their counterparts in the aluminium-magnesium-oxygen-water system can be calculated using equations derived in Appendix 2. Some values for a range of temperatures relevant to production processes are given in Table 8.6.

The free energies of formation for the oxide species of the alloying elements are all more negative than that for corundum, so that if the lithium activity in an aluminium lithium alloy is sufficient, the oxidation product expected in *dry air* is lithium oxide or the spinel.

However, for the normal range of water vapour pressures in *humid air*, lithium hydroxide is stable with respect to lithium oxide, as illustrated in Table 8.4 for the water vapour pressure range, 500 to 4000 Pa, and the temperature range 300 to 600°C. This represents a significant difference from the oxidation of aluminium-magnesium alloys. It suggests that lithium hydroxide is a primary oxidation product on aluminium-lithium alloys, enabling the hydrogen potential of the atmosphere to be applied to the metal surface through a continuous path for hydroxyl ions, promoting hydrogen absorption.

8.4.3.2 Hydrogen Absorption in Aluminium Lithium Alloys from Clean Humid Air

The expected sensitivity of aluminium-lithium alloys to hydrogen absorption from humid air was confirmed by Henry[49] who provided evidence from experiments conducted for an aluminium- 2 mass% lithium binary alloy and the AA 2091 alloy which also contains 2 mass% of lithium. The method was to measure the uptake of hydrogen in freshly prepared clean samples heated isothermally in air with controlled humidity. The samples used were 10 mm diameter × 60 mm cylinders with a fine surface finish dry machined from rod rolled to 75% reduction from cast material.

Heat Treatment

Several samples were maintained at 500°C in flowing clean air with a water vapour partial pressure of 3050 Pa (0.03 atm.) introduced by saturating it at the appropriate dew point, 25°C. Samples were withdrawn successively at intervals of 1 hour throughout a period of 8 hours.

Determination of Hydrogen Content

The hydrogen contents of the heat-treated samples were determined by vacuum extraction, taking the strict precautions for alloys with active volatile components described in Chapter 5, Section 5.2.3.6.

Preliminary experiments revealed that hydrogen absorbed in the treatments was located very close to the surface. In preparing samples for determinations, sufficient material must be removed from the surface of the samples to avoid the generation of spurious hydrogen as explained in Chapter 5, Section 5.2.3.1 but some of the subcutaneous zone containing the hydrogen absorbed is inevitably lost. With care and practice, the metal surface could be skimmed on a lathe, removing the whole of the geometric surface and yet losing only 5 μm of metal from the surface. Even so, eccentricity of as little as 2 μm introduced significant variability in the quantity of metal removed from the surface and hence variability in the results, which had to be accepted.

Quantities of Hydrogen Absorbed

The results of hydrogen content determinations are plotted as the solid circles in Figures 8.19 and 8.20 for the experimental 2 mass% lithium binary alloy and the commercial alloy AA 2091 respectively. The quantities of hydrogen absorbed were exceptionally high and the similarity of results for the different alloys show that the effect is general.

8.4.3.3 Lithium Hydride as a Sink for Absorbed Hydrogen

Micrographic observations revealed that the hydrogen was trapped as lithium hydride precipitated just below the metal surface. Cross-sections of the samples remote from the ends were mounted in bakelite under pressure, taking care to support and preserve the edges. They were polished for microexamination on water-lubricated graded silicon carbide papers, finishing with automatic polishing on 6 μm diamond paste and then on a colloidal silica suspension in a flood of a dry unreactive light oil lubricant on soft napped cloth to minimise corrosion of lithium hydride.

Figure 8.21 illustrates lithium hydride particles observed in the size range 1 to 4 μm, copiously distributed in a subcutaneous zone approximately 150 μm wide with shapes expected from random sections through cubic crystals of lithium hydride as illustrated at higher magnification in Figure 8.22, e.g. squares, rectangles and triangles. There is a thin hydride free band about 40 μm wide, between the surface and the precipitate zone due to preferential nucleation of hydride on the surface. The morphology of the particles is similar to that of sodium hydride dispersed in an aluminium matrix illustrated in Figure 4.5, as expected from the similar NaCl type structure of the two hydrides. The precipitation of hydrogen as hydride on entering the metal renders pre-existing pore nuclei irrelevant as a sink for absorbed hydrogen. In this respect, the location of absorbed hydrogen in

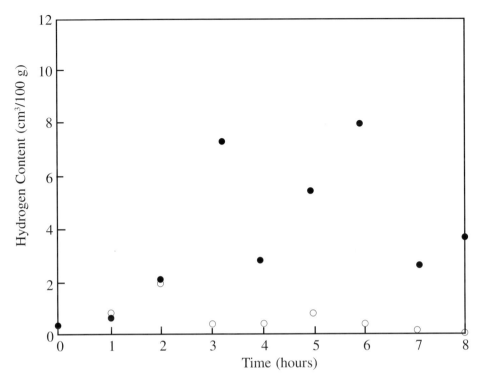

Fig. 8.19 Hydrogen absorbed by an aluminium - 2 mass% lithium binary alloy at 500°C in (a) humid air and (b) humid air with added sulphur dioxide.

●   air with 3050 Pa (0.03 atm.) water vapour partial pressure.
○   air with 3050 Pa (0.03 atm.) water vapour partial pressure and 1000 Pa (0.01 atm.) sulphur dioxide partial pressure.

aluminium–lithium alloys differs fundamentally from the accommodation generated by the expansion of pores in aluminium–magnesium alloys, described earlier.

The local hydrogen content of the subcutaneous metal can be estimated from the size and distribution of the hydride particles. Assume, for example, that lithium hydride is present as 1 μm cubes with an average spacing of 10 μm. The volume of every particle is 10^{-12} cm^3 and the density of LiH is 0.8 g/cm^3, so that the mass of the particle is 8×10^{-13} g and contains $\sim 10^{-13}$ moles of LiH. One mole of LiH requires ½ mole of hydrogen for its formation and the density of the alloy is approximately 2.8 g/cm^3. Hence its contribution to the hydrogen content of the surrounding metal volume of 10^{-9} cm^3 is:

$$\frac{10^{-13} \times 22400 \times 100}{2 \times 2.8 \times 10^{-9}} \text{ cm}^3/100\text{g} = 43 \text{ cm}^3/100\text{g} \tag{8.35}$$

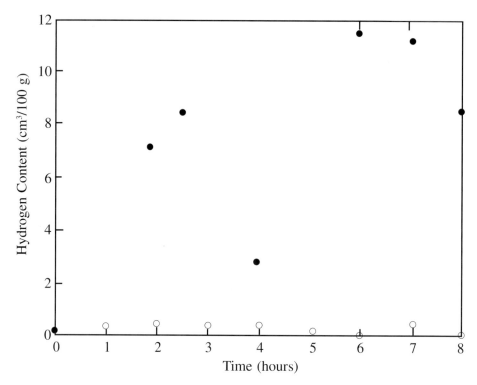

Fig. 8.20 Hydrogen absorbed by AA 2091 (2% Li, 2% Cu, 1.5 % Mg, 0.12 Zr) at 500°C in (a) humid air and (b) humid air with added sulphur dioxide.
● air with 3050 Pa (0.03 atm.) water vapour partial pressure.
○ air with 3050 Pa (0.03 atm.) water vapour partial pressure and 1000 Pa (0.01 atm.) sulphur dioxide partial pressure.

8.4.3.4 The Influence of Sulphur Pollution in the Atmosphere

In view of the marked effect of sulphur pollution on the absorption of hydrogen by aluminium-magnesium alloys, Henry[49] repeated the experiments described in Section 8.4.3.2 using an atmosphere of with the same humidity but modified by the addition of 1000 Pa (0.01 atm) of sulphur dioxide. The results of the hydrogen content determinations are plotted as the open circles in Figures 8.19 and 8.20. The hydrogen contents after the heat treatments are very low and microexamination of cross sections through the samples failed to disclose any trace of hydride precipitation. Paradoxically, whereas sulphur dioxide stimulates the absorption of hydrogen by aluminium-magnesium alloys it inhibits absorption by aluminium-lithium alloys.

There is insufficient information on the oxidation kinetics of solid aluminium-lithium alloys to predict the constitution of the oxidation products expected when the system is further complicated by the introduction of sulphur chemistry with possibilities of sulphides

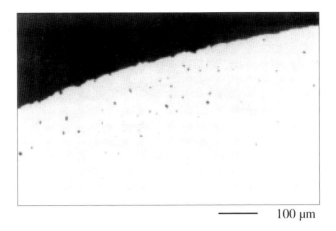

—— 100 μm

Fig. 8.21 Hydride particles in a subcutaneous zone of an aluminium-lithium alloy heated for 4 hours in air with water vapour pressure of 3050 Pa (0.03 atm.). Mechanically polished microsection.

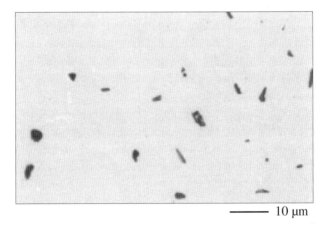

—— 10 μm

Fig. 8.22 As Figure 8.21 at higher magnification, showing shapes of particles consistent with random sections through cubic crystals.

and sulphates as components. The inhibition of hydrogen absorption is undoubtedly due to a change in the oxidation mechanism that blocks the transport of hydroxyl ions. Among various factors to be considered is the alkalinity of lithium hydroxide and the exceptionally high solvation energy of the Li^+ ion which persists in its salts,[48] especially Li_2SO_4. Thus the conversion of the alkaline hydroxide to lithium sulphate:

$$4LiOH + 2SO_2 + O_2 = 2Li_2SO_4 + 2H_2O \qquad (8.36)$$

could both eliminate the pathway for hydroxyl ions and immobilise water:

$$(Li^+)_2SO_4 + 8H_2O = (Li^+.4H_2O)_2SO_4 \qquad (8.37)$$

It is an interesting effect that will no doubt be addressed in future.

8.5 HYDROGEN ABSORBED FROM ROLLED SURFACES

8.5.1 CHARACTERISTICS OF HOT-ROLLED SURFACES

Sheet products are usually produced by hot-rolling preheated ingots with multiple passes to an intermediate material with a thickness in the range 4 to 12 mm, which is then cold rolled to the final gauge. The hot-rolled intermediate material has various producer specific designations but in the present context it is convenient to refer to it as *plate* although the term is usually applied to thicker finished products produced by hot-rolling only.

During hot-rolling, deformation in the roll gap introduces differential velocities between the stock and the roll, so that the stock slips backwards on the roll face on entry and forwards on exit. The relative motion induces transfer of aluminium particles to the roll face which are carried round and transferred back to the metal as the roll revolves. As successive ingots are rolled, a steady state is established in which the roll surface carries a constant thin coating of the metal and the surface of the emerging metal stock is a thin layer of compacted flake. The roll is lubricated and cooled with a recirculated soluble oil/water emulsion, which interacts with the metal surface yielding oxides and metal soaps, degrading the lubricant.

The topography of a representative hot-rolled surface of AA 1100 pure aluminium, exhibiting flakes typically 0.20×0.05 mm, interleaved in the rolling direction is illustrated at high magnification in the photomicrograph given in Figure 8.23. The corresponding section through it, given in Figure 8.24, reveals that the flakes are compacted by rolling to form a subcutaneous zone on the metal, about 6 μm thick, entrapping extraneous matter dispersed as small, discrete inclusions. This material contains water bearing species, e.g. hydrated aluminas and metal soaps derived from reaction with the lubricant.

8.5.2 PROFILE OF THE SURFACE HYDROGEN SOURCE

The profile of the subcutaneous zone as a hydrogen source can be explored by the following laboratory assessment. If a piece of hot-rolled aluminium is heated to a sufficient temperature in a high vacuum system from which gas evolved is continuously pumped into low pressure system, as e.g. described in Section 5.2 of Chapter 5, the gas collected is the sum of the actual hydrogen content of the metal and the contribution from the surface source, reacting with the metal. Thus if samples with progressively greater depths

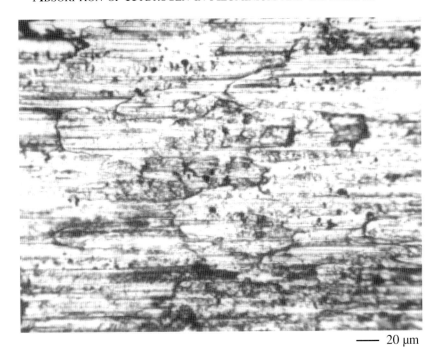

—— 20 μm

Fig. 8.23 Surface topography of 6 mm thick hot-rolled AA 1100 aluminium plate, showing compacted interleaved flakes.

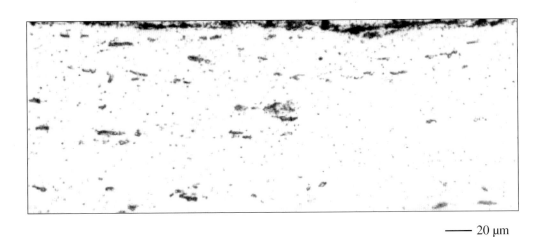

—— 20 μm

Fig. 8.24 Section through the hot-rolled aluminium plate in Figure 8.23, showing extraneous material trapped between the flakes.

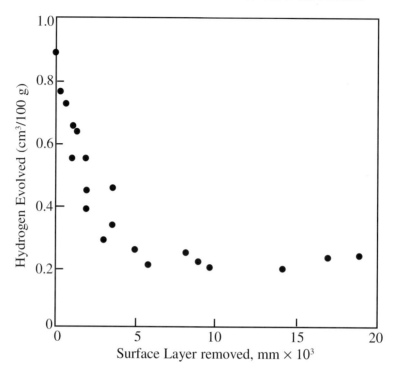

Fig. 8.25 Hydrogen evolved in vacuum from hot-rolled 6.35 mm thick AA 1100 aluminium plate as a function of depth of surface removed.

of metal removed from their surfaces are submitted to the test, the effective depth of the source corresponds with the minimum depth of surface that must be removed to yield the true hydrogen content of the metal. To avoid distorting the surface, the thin layer of metal is removed chemically, e.g. etched in sodium hydroxide solution and rinsed in dilute nitric acid and distilled water.

Figure 8.25 records the results of such measurements at 500°C on a representative piece of hot rolled AA 1100 aluminium, 6.35 mm thick with an actual hydrogen content of 0.24 ± 0.04 cm^3/100 g. The additional quantity of hydrogen due to the surface source diminished progressively as more and more of the surface was removed until it was reduced to zero when 6 μm of metal had been removed, corresponding with the depth of the subcutaneous zone illustrated in Figure 8.24. The inference is that the metal retains an integral surface source of hydrogen as long as any of the implanted extraneous material remains.

Figure 8.26 records corresponding measurements for the same material after it had been heated to 550°C at the rate of 3.0×10^{-2} K s^{-1} to allow the metal to react with the surface contamination. The latent hydrogen associated with the surface source was then

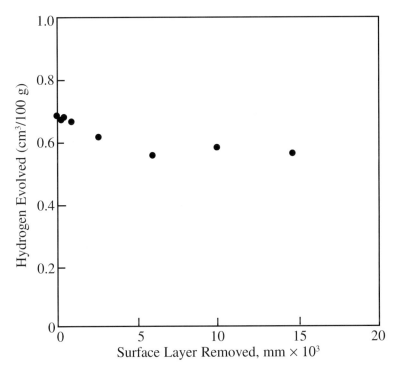

Fig. 8.26 Hydrogen evolved in vacuum from 6.35 mm thick hot-rolled AA 1100 aluminium plate as a function of depth of surface removed after previously heating it from 20 to 550°C with the temperature rising at 3×10^{-2} K s^{-1}.

evolved irrespective of whether the subcutaneous zone was removed or not, so that it had been converted to actual hydrogen in solution. The form of Figure 8.26 and its relation to Figure 8.25 are consistent with the imposition of a constant hydrogen potential at the metal surface, corresponding to a hydrogen content of about 0.6 cm^3/100 g. This view is logical because the hydrogen potential is associated with the free energy change and kinetics of reaction between the metal and hydrous aluminas implanted in the subcutaneous zone. The information is applied later in Section 8.5.3.2 to define an assumed boundary condition in a mathematical model for the hydrogen absorption.

8.5.3 HYDROGEN ABSORPTION IN ANNEALING DURING MANUFACTURE

Ingots are preheated to typically 500°C for hot-rolling but successive chills against the roll face reduces the temperature of the metal to below that for continuous recovery so that the nominally hot-rolled metal is partially cold-worked. If it is to be cold-rolled, the hot-rolled metal is annealed in electric furnaces to soften it. This provides the first opportunity for latent hydrogen in the surface source to enter the metal.

In industrial batch annealing, time and temperature are inseparable because of the thermal inertia of a large mass of cold metal comprising a charge. For example, if a representative charge of 20 tonnes is to be heated for a nominal period of twenty minutes at 400°C, it may take the metal as long as 4 hours to attain the final temperature, eliciting:
1. Hydrogen absorption over an extensive temperature range.
2. Possible dehydration of the extraneous material implanted in the surface.

Correct representation of industrial conditions must therefore take account of the progressively rising temperature.

8.5.3.1 Simulated Annealing of Hot Rolled Aluminium
Figure 8.27 records the hydrogen contents determined by vacuum extraction on samples taken through the full thickness of a 6.35 mm thick hot-rolled aluminium plate after simulated industrial heat treatments in air. The temperature was raised progressively at 3.0×10^{-2} °C s^{-1} to match the thermal response of a 20 tonne charge, which typically requires 3 to 4 hours to reach 400°C from ambient temperature, e.g. 20°C. A further factor for assessment is whether natural humidity in the normal atmosphere can influence the activity of the surface hydrogen source, so the information is given for two atmospheres with different water vapour partial pressures:
1. 3000 Pa (0.03 atm.) representing high humidity in ambient air.
2. $< 5 \times 10^{-2}$ Pa (5×10^{-7} atm.) representing very dry air.

As a control, corresponding results are given for the same plate heated in air with the high humidity after removing the subcutaneous zone of compacted flake by etching.

Figure 8.27 confirms that the contaminated subcutaneous zone is the primary source of the hydrogen and it is eliminated when the zone is removed. Atmospheric humidity has only a minor influence and the effect is insufficient to justify detailed consideration. The somewhat lower hydrogen contents for metal heated in very dry air is doubtless due to slight dehydration of the surface.

The entry of hydrogen is diffusion-controlled and, as expected, it is insignificant at temperatures < 350°C but thereafter it rises rapidly as the temperature increases further. There is only a narrow temperature range between the minimum of about 380°C needed to soften the metal and a maximum of about 400°C above which hydrogen absorption is rapid.

8.5.3.2 Mathematical Model of Hydrogen Absorption
The simplest view is that the hydrogen source sets a constant surface hydrogen potential, as indicated in Section 8.5.2. With this assumption, the absorption of hydrogen can be calculated, using eqn 7.7, solved graphically with the use of Figure 7.3, given in Chapter 7, from which the value of Q_t/Q_∞ can be obtained as a function of the dimensionless parameter, Dt/a^2:

$$Q_t/Q_\infty = (c_m - c_o)/(c_s - c_o) \tag{8.38}$$

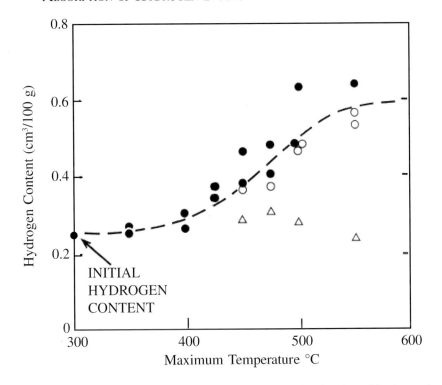

Fig. 8.27 Quantity of hydrogen absorbed from contamination implanted in the surface of 6.35 mm thick AA 1100 (99.2% aluminium) plate by hot-rolling.

Samples heated in air with temperature rising by 3.0×10^{-2} K s^{-1} from 20°C, simulating the thermal response of a typical charge in an industrial furnace.

- ●   Untreated hot-rolled plate. Water vapour pressure: 3000 Pa
- ○   Untreated hot-rolled plate. Water vapour pressure: $< 5 \times 10^{-2}$ Pa.
- △   Plate etched to remove contamination. Water vapour pressure: 3000 Pa.
- $---$ Locus of values calculated using diffusion theory.

where:

c_o is the initial hydrogen concentration in the plate, assumed uniform.

c_s is the surface concentration, assumed constant.

c_m is the mean concentration after the elapse of time, t.

    In the present problem, the diffusion coefficient, D, is not constant but rises progressively as the metal temperature rises from its initial value, T_I, to its final value, T_F, by the Arrhenius' type equation, given as eqn 7.2 in Chapter 7:

$$D = D_o \exp \frac{-\Delta G^*}{RT} \tag{8.39}$$

The problem can be approached using the standard method for a time-dependent diffusion coefficient, in which the function $(Dt)dt$ is integrated over the period of interest.[50, 51] To apply it to the present problem, the temperature dependence of D is converted to time dependence by substituting for T in eqn 8.39 from the linear equation describing the rise in temperature:

$$T = T_I + kt \qquad (8.40)$$

yielding:

$$D = D_o \exp \frac{-\Delta G^*}{R(T_I + kt)} \qquad (8.41)$$

Hence:

$$\frac{Dt}{a^2} = \frac{1}{a^2} \int (Dt)dt = \frac{D_o}{a^2} \int \exp \frac{-\Delta G^*}{R(T_o + kt)} dt \qquad (8.42)$$

Integrating between the limits, $t = 0$ and $t = t_F$ yields:

$$\frac{Dt}{a^2} = \frac{D_o \Delta G^*}{ka^2 R} \left[\frac{e^{-y}}{y^2} \left\{ 1 - \frac{2!}{y} + \frac{3!}{y^2} - ... \right\} - \frac{e^{-x}}{x^2} \left\{ 1 - \frac{2!}{x} + \frac{3!}{x^2} - ... \right\} \right] \qquad (8.43)$$

where $x = \Delta G^*/RT$, $y = \Delta G^*/RT_F$ and T_F is the final temperature after a time, t_F.

The model represented by eqn 8.43 can be used to confirm the nature assumed for the surface hydrogen source by applying it to the empirical results described in Section 8.5.3.1. Putting in the numerical values, $k = 3.0 \times 10^{-2}$ K s^{-1}, $a = 3.17$ mm and values of $\Delta G^*/RT$ and D_o obtained by matching eqn 7.2 with eqn 7.27 in Chapter 7 (representing diffusion in a field of traps characteristic of manufactured material) yields the values of Dt/a^2 for selected final temperatures, given in Table 8.7. The values of $(c_m - c_o)/(c_s - c_o)$ are identical with the corresponding values of Q_t/Q_∞ which can be read from Figure 7.3. Using the value $c_s = 0.60$ cm^3/100 g for the surface concentration corresponding to the constant surface hydrogen potential suggested by Figures 8.25 and 8.26, and the measured value for the initial concentration, $c_o = 0.25$ cm^3/100 g gives the mean hydrogen contents in the fourth column of the table. These calculated values correlate well with the results of the practical determinations given in the fifth column of the table and plotted in Figure 8.27.

The development of the hydrogen concentration profile through the plate half-thickness can be assessed by using the values of Dt/a^2 given in Table 8.7 to select plots of $(c - c_o)/(c_s - c_o)$ against x/a in Figure 7.1 to suit particular thermal histories. This indicates significant enrichment in hydrogen content even at temperatures for which the mean concentration in the material is quite low. For example, when the metal temperature has reached 400°C, Table 8.7 gives $Dt/a^2 = 0.011$ and selecting the plot for this value in Figure 7.1 yields values for $(c - c_o)/(c_s - c_o) > 0.46$ at any plane within a 0.1 fraction of

Table 8.7 Values of Dt/a^2 and $(c_m - c_o)/(c_s - c_o)$ as Functions of Temperature for Diffusion of Hydrogen into Plate Heated from 20°C at 3×10^{-2} K s^{-1}

Final Temperature°C	Dt/a^2	$\dfrac{c_m - c_o}{c_s - c_o}$	Hydrogen Content (cm^3/100 g)	
			Calculated[†]	Actual
550	1.68	0.98	0.60	0.64
525	0.89	0.91	0.57	0.48, 0.63
500	0.40	0.70	0.50	0.48
450	0.083	0.32	0.36	0.37, 0.46
400	0.011	0.10	0.29	0.30, 0.26
350	0.00014	0	0.25	0.26, 0.26

[*]Read from Figure 7.3 as Q_t/Q_∞

C_o = Initial concentration, assumed uniform, taken as 0.60 cm^2/100 g.
C_s = Surface concentration, assumed constant, taken as 0.25 cm^2/100 g.
C_m = Mean concentration after elapse of time, t.
D = Diffusion coefficient.
a = Plate half thickness.

the thickness below the metal surfaces. For $c_o = 0.25$ and $c_s = 0.60$ as before, when the metal temperature has reached 400°C, the local hydrogen concentration in the surface zone is > 0.41 cm^3/100 g, while Table 8.7 gives the mean concentration through the whole thickness as only 0.29 cm^3/100 g. The high subcutaneous hydrogen content is particularly deleterious if residues of near surface porosity are inherited from unsatisfactory ingots, as discussed later in Chapter 10.

8.5.3.3 Effect of Cold Rolling

Subsequent cold-rolling further compacts and burnishes the flake surface produced by hot-rolling and also impresses its own characteristic artefact on the surface, i.e. reticulation, a network of very fine, shallow surface cracks, offering traps for contamination from the mineral oil lubricants used. Figure 8.28 shows that cold rolling raises the potential of the hot rolled surface hydrogen source if is left intact. However, as Figure 8.28 also shows, removing it by etching the hot rolled metal before cold rolling it does not completely eliminate the risk of hydrogen absorption during subsequent annealing because cold rolling introduces new, albeit less serious surface contamination.

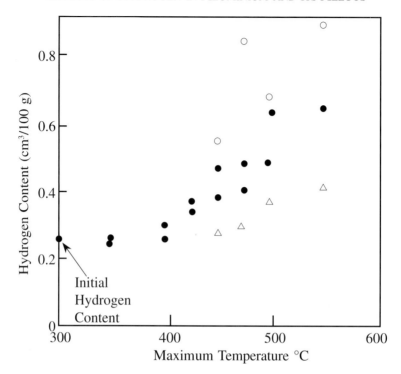

Fig. 8.28 Effect of subsequent cold-rolling on the quantity of hydrogen absorbed from surface contamination implanted by hot-rolling.

Samples of 3.17 mm thick AA 1100 (99.2%) aluminium sheet cold-rolled from 6.35 mm hot-rolled plate, heated in air with temperature rising from 20°C by 3.0×10^{-2} K s^{-1}. Water vapour partial pressure: 3000 Pa.

○   Sheet cold-rolled from untreated hot-rolled plate.

△   Sheet cold-rolled from hot-rolled plate etched to remove contamination.

●   Untreated hot-rolled plate (extracted from Figure 8.27 For comparison).

8.6 REFERENCES

1.   O. Kubaschewski and C.B. Alcock, *Metallurgical Thermochemistry*, Pergamon Press, New York, 1979.
2.   D.E.J. Talbot, *International Met. Reviews*, **20**, 1975, 166.
3.   C.E. Ransley and D.E.J. Talbot, *Z. Metallkunde*, **46**, 1955, 328.
4.   H. Kostron, *Z. Metallkunde*, **43**, 1952, 269 and 373.
5.   O. Kubaschewski, A. Cibula and D.C. Moore, *Gases and Metals*, London, 1970, (Iliffe).
6.   L.W. Eastwood, *Gas in Light Alloys*, Chapman and Hall, London, 1946.
7.   L Sokol'skaya, *Gas in Light Metals*, Pergamon Press, London, 1961.

8. E.F. Emley and A.V. Brant, *Neue Verfahren fur die Halbzeugherstellung*, Symp. Deutsche Gezellschaft fur Metallkunde, Bad Homburg, 1973.

9. D. Hanson and I.G. Slater, *J. Inst. Metals*, **46**, 1931, 216.

10. R. Eborall and C.E. Ransley, *J. Inst. Metals*, **71**, 1945, 525.

11. W.R. Opie and N.J. Grant, *The Foundry*, **78**, 1950, 104.

12. D. Hanson and I.G. Slater, *J. Inst. Metals*, **46**, 1931, 187.

13. A.J. Swain, *J. Inst. Metals*, **80**, 1951–52, 125.

14. A.A. Gorshkov and V.S. Vargin, *Liteinoe Proizv.*, **8**, 1954.

15. K. Wefers and G.M. Bell, Technical Paper No. 19, Alcoa Technical Center, Pa 15069, USA.

16. D.E.J. Talbot and P. N. Anyalebechi, *Materials Science and Technology*, **4**, 1988, 1.

17. C. Wagner, *Z. Phys Chem.* B, **21**, 1933, 25.

18. R.E.Newham and Y.M. DeHaan, *Z. Krystallogr.*, **117**, 1962, 235.

19. W.D. Kingery and G.E. Meiling, *J. Appl. Phys.*, **32**, 1961, 556.

20. T. Matsumura, *Can. J. Phys.*, **44**, 1966, 1685.

21. J. Anderson, Discussion of Paper by J.H. de Boer and G.M.M. Houben, *Proc. Inter. Symp. Reactivity of Solids*, Gothenburg, 1952.

22. H.J.T. Ellingham, *J. Soc. Chem. Ind.*, **63**, 1944, 125.

23. D.J. Stephenson, 'The Absorption of Hydrogen from Humid Atmospheres by Molten Aluminium and an Aluminium-Magnesium Alloy', Ph.D. Thesis, Brunel University, 1978.

24. R. Dholiwar, 'The Oxidation of Liquid Aluminium', Ph.D. Thesis, Brunel University, 1983.

25. C. Lea and D.J. Ball, *Appl. Surf. Sci.*, **17**, 1984, 344.

26. I.M. Ritchie, J.V. Saunders and P.L. Weiekhard, *Oxid. Metals*, **3**, 1971, 91.

27. A.J. Brock and M.A. Heine, *J. Electrochem. Soc.*, **119**, 1972, 1123.

28. G.M. Scamens and E.P. Butler, *Met. Trans. A*, **6**, 1975, 2055.

29. M.P. Silva, 'Oxidation of Aluminium-Magnesium Alloys in the Solid, Semi-Liquid and Liquid States', Ph.D. Thesis, Brunel University, 1987.

30. M.P. Silva and D.E.J. Talbot, *Light Metals*, 1989, 1035.

31. C. Renon and J. Calvet, *Mem. Sci. Rev. Met.*, **58**, 1961, 835.

32. C. Renon and J. Calvet, *Mem. Sci. Rev. Met.*, **60**, 1963, 620.

33. D.E.J. Talbot and D.A. Granger, *J. Inst. Metals*, **92**, 1963–4, 290.

34. J.M.Morton, 'Hydrogen Absorption and Loss During Heat treatment of Aluminium Alloys Containing Magnesium and Lithium', Ph. D. Thesis, Brunel University, 1989.

35. S.P. Mitoff, *J. Chem. Phys.*, **36**, 1962, 1383.

36. M.O. Davies, *J. Chem. Phys.*, **38**, 1963, 2047.

37. H. Schmaltzried, *J. Chem. Phys.*, **33**, 1960, 940.

38. G.H. Reiling and E.B. Hensley, *Phys. Rev.*, **112**, 1958, 1106.

39. D.R. Sempolinski, W.D. Kingery and L. Tuller, *J. Amer. Cer. Soc.*, **63**, 1980, 669.

40. A.M. Glass and T.M. Searle, *J. Chem. Phys.*, **46**, 1967, 2092.

41. H.B. Johnsen, O.W. Johnsen and I.B. Cutler, *J. Amer. Cer. Soc.*, **49**, 1966, 390.

42. K.J. Kurzydlowski and B. Ralph, *Quantitative Description of the Microstructure of*

Materials, CRC Press, Boca Raton, 1995.

43. O. Kubaschewski and B.E. Hopkins, *Oxidation of Metals and Alloys*, London, Butterworth, 1962.

44. P.T. Stroup, US Patents Nos. 2092033 and 2092034.

45. C. Smith and N. Swindells, *J. Inst. Metals*, **82**, 1953–4, 323.

46. R. Tahbaz, 'Absorption of Hydrogen by a Solid Aluminium-Magnesium Alloy from Humid Atmospheres', M. Phil. Thesis, Brunel University, 1977.

47. *Handbook of Chemisty and Physics*, Boca Raton Fla, CRC Press, 1997.

48. F.A. Cotton and G. Wilkinson, *Advanced Inorganic Chemistry*, Interscience Division, John & Wiley Sons, New York, 1966.

49. D.M.Henry, 'The Nature and Effects of Hydrogen in Weldalite Aerospace Alloy and Other Aluminium-Lithium Alloys', Ph.D. Thesis, Brunel University, 1995.

50. J. Crank, *The Mathematics of Diffusion*, Clarendon Press, Oxford, 1956.

51. H.S. Carslaw and J.C. Jaeger, *Conduction of Heat in Solids*, Oxford University Press, Oxford, 1947.

9. Porosity and Related Artifacts in Manufactured Products

When liquid aluminium or any of its common alloys is cast by the techniques used in normal production routes, it retains virtually all of the dissolved hydrogen it may contain[1, 2] as illustrated in Figure 9.1, but because the solubility is less in the solid than in the liquid metal, some of the gas is rejected from solution and is entrapped as an internal gas phase in the structure, generating one or both of two different kinds of porosity:

1. *Interdendritic (or primary) porosity* generated by hydrogen rejected from the interdendritic liquid fraction of the solidifying metal as irregular voids, often interconnected, easily resolved by a low power microscope and often visible to the unaided eye.[1–4]

2. *Secondary porosity*[5] nucleated immediately after solidification as innumerable small isolated spherical voids, typically 1 or 2 μm in diameter.

The current chapter deals with the principles underlying the generation of these two kinds of porosity and their persistence through thermal and mechanical treatments to intermediate and final products. The information is derived partly from experiment and partly from observations of metal in production. Its implications for controlling the quality of manufactured products are considered in Chapter 10.

9.1 EVALUATION OF POROSITY

The principle methods used to acquire reliable information on porosity are quantitative measurement of porosity volume fractions, examination of pores in microsections and the determination of hydrogen contents.

Measurement of Porosity Volumes
Porosity measurements are required for the range 0 to 1% by volume, with a sensitivity of ± 0.02%. The methods must be suitable for both closed and open pore systems.

Porosity volumes can be calculated from the densities of samples of the metal before and after consolidating them by applying pressure at elevated temperature in a closed die. It is customary to express the values as % porosity. The samples must be large enough to meet the required sensitivity but small enough to permit closely spaced sampling from standard metal products and intermediates. It is usually convenient to use cylindrical samples 2 cm in diameter × 1.5 cm high.

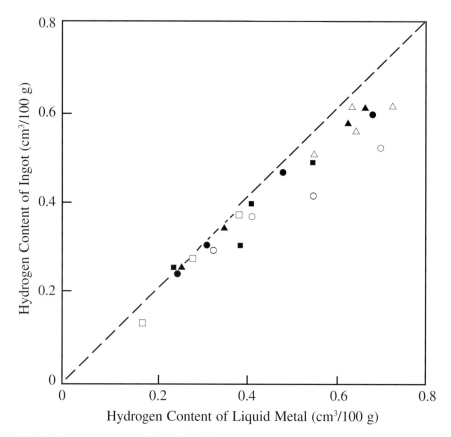

Fig. 9.1 Hydrogen contents of small scale experimental melts and corresponding ingots demonstrating retention of hydrogen in semicontinuous casting. Cross-section of ingots 15 × 3.75 cm. (Ransley and Talbot).[2]

△   99.99% pure aluminium cast at 0.21 cm s⁻¹.

○   99.8% pure aluminium cast at 0.21 cm s⁻¹.

●   99.2% pure aluminium cast at 0.21 cm s⁻¹.

▲   99.2% pure aluminium cast at 0.42 cm s⁻¹.

■   99.2% pure aluminium cast at 0.63 cm s⁻¹.

□   Duralumin type alloy (AA 2014) cast at 0.21 cm s⁻¹.

A suitable consolidation procedure is to extract the hydrogen supporting the pores by heating the samples in vacuum and then to apply a pressure of 150 MPa to them at 450°C in a closed die for 15 minutes. Corrections are applied for possible error due to phase changes during consolidation by conducting blank determinations in which reference samples are subjected to the same thermal cycle but without applying pressure.

The densities are measured by Archimedes method in which samples are weighed in air and suspended in water. To satisfy the required sensitivity, it is necessary to eliminate errors from air dissolved in the water, surface tension forces on the suspension wire and the temperature-dependence of the densities of both water and the metal. Using these procedures, errors can be restricted to ± 0.01% porosity.

Preparation of Microsections

Interdendritic porosity can be clearly resolved in microsections polished by any of the standard methods and it can usually be resolved in macrosections by fine machining.

Secondary porosity is usually visible as very small circular features in microsections polished by standard procedures on lubricated emery papers and graded diamond pastes to a 1 μm finish but because aluminium is soft and some of the pores are of comparable size to the abrasives, e.g. 2 μm in diameter it is difficult to avoid distorting them by burring over their unsupported edges. Electropolishing in 1 : 5 perchloric acid/ethyl alcohol solution for a short time is usually needed to delineate the pores clearly for micrographs. Quantitative metallography[6] on electropolished microsections, yields values for porosity proportional to corresponding determinations from density measurements but overestimates them by a factor of at least 2.

Measurement of Hydrogen Contents

Hydrogen contents are determined by the vacuum extraction method described in Chapter 5, Section 5.2. Samples of solid metal are prepared by machining as described in Chapter 5, Section 5.2.3.1. Samples of liquid metal are machined from well fed test bars chill cast from the metal stream where it enters the mould at the position of interest in the subsequent solid ingot or casting. Samples with flaws are discarded and replaced.

9.2 INTERDENDRITIC POROSITY

9.2.1 Theory of Formation

As an ingot or a casting solidifies some but not all of the hydrogen it contains is transferred from the growing solid crystals into the adjacent liquid as a consequence of the difference in solubility between the solid and liquid metal.[1-4, 7] The hydrogen content in the diminishing liquid fraction rises until it locally exceeds a value sufficient to nucleate the gas phase, i.e. a value for which the equilibrium hydrogen pressure exceeds the sum of the external pressure acting on the liquid, the pressure due to the head of liquid metal and the excess pressure required to oppose surface tension. Since this condition is usually reached near the end of solidification, the gas bubbles are constrained to occupy the restricted spaces between the growing dendrites, so that they become irregular gas-filled cavities, with the characteristic shapes illustrated in Figure 9.2.

The quantity of porosity so formed depends not only on the hydrogen content of the liquid metal but also on the rate of abstraction of heat and on the freezing range of the solidifying metal. The influence of these factors has been assessed from the quantitative

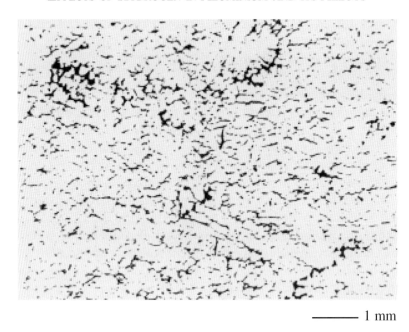

——— 1 mm

Fig. 9.2 Extensive interdendritic porosity in 99.2% aluminium (AA 1100) showing characteristic irregular shapes and interconnected distribution. Vertical section through a semicontinuously cast ingot. Hydrogen content 0.25 cm³/100 g.

results reproduced in Figures 9.3 and 9.4, taken from various sources.[1-4, 7] The significant features are as follows:

1. Gross porosity occurs only if the hydrogen content exceeds certain critical values appropriate to the casting conditions and the composition of the metal.
2. Compared with sand casting, the more rapid cooling in semi-continuous casting suppresses the formation of porosity to a degree determined by the freezing range of the metal. Less and less interdendritic porosity is formed as the freezing range diminishes.
3. Plots of porosity against hydrogen content are either straight lines or are asymptotic to lines which have slopes consistent with the view that the hydrogen pressure in the pores is only a little higher than atmospheric pressure at the time of their formation.[1, 2, 7]

Ransley and his collaborators[1, 2] explained these effects on the assumption that the hydrogen is initially co-deposited with the solid metal and is then rejected into the adjacent liquid by diffusion across the boundaries of the growing dendrites. The quantity of porosity formed is determined by factors controlling the transfer of hydrogen into the liquid, especially the cooling rate and freezing range which together determine the time for which the solid and liquid fractions are in short range contact and the area of contact between them.

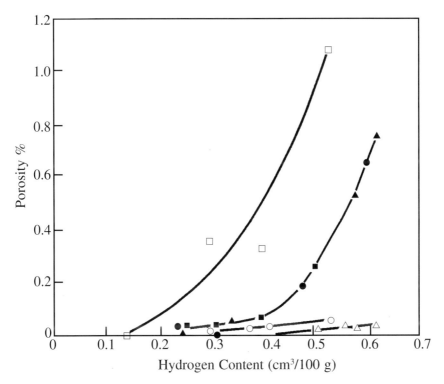

Fig. 9.3 Relation between hydrogen content and porosity for small-scale experimental semicontinuously cast ingots of aluminium of various purities and a Duralumin type alloy (AA 2014), showing the effect of freezing range. Cross-section of ingots 15 × 3.75 cm. (Ransley and Talbot).[2]

△ 99.99% pure aluminium cast at 0.21 cm s^{-1}.

○ 99.8% pure aluminium cast at 0.21 cm s^{-1}.

● 99.2% pure aluminium cast at 0.21 cm s^{-1}.

▲ 99.2% pure aluminium cast at 0.42 cm s^{-1}.

■ 99.2% pure aluminium cast at 0.63 cm s^{-1}.

□ Duralumin type alloy (AA 2014) cast at 0.21 cm s^{-1}.

This concept can account both for the insensitivity of very pure aluminium to hydrogen porosity and for the critical values of hydrogen content below which interdendritic porosity does not form in lower purity metal and many of the common alloys, illustrated in Figures 9.3 and 9.4. Very pure aluminium has virtually no freezing range and therefore little opportunity for hydrogen to transfer from the solid into the liquid so that porosity is suppressed. The same applies to the last liquid fractions of lower purity grades of aluminium and many of the common alloys because they approach eutectic compositions. Such a composition also has virtually no freezing range, so that porosity corresponding to the hydrogen concentrated into it is frozen into the solid in supersaturated solution

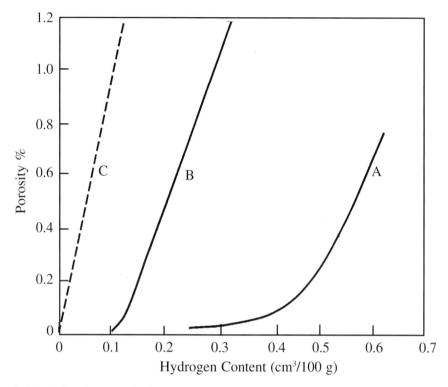

Fig. 9.4 Relation between hydrogen content and porosity for small-scale experimental semicontinuously cast ingots and sand cast bars of 99.2% pure aluminium and comparison with theoretical limiting porosity.
Curve A - Semi-continuously cast ingots (extracted from Figure 9.3).
Curve B - 2.5 cm diameter sand cast test bars (Ransley and Neufeld).[1]
Curve C - Limiting porosity equivalent to the volume of diatomic hydrogen if released within the metal at 660°C and atmospheric pressure.

and the porosity that would otherwise correspond to it is suppressed. A critical value of hydrogen content is thus interpreted as the quantity that just fails to saturate the final liquid fraction when concentrated into it as described. Critical values are typically in the range 0.10 to 0.15 cm³/100 g depending on the metal composition and casting parameters as illustrated in Figures 9.3 and 9.4.

In practice, the final liquid fractions of commercial grades of aluminium and its alloys are not ideal eutectic compositions and solidify over a narrow temperature range. Consequently, a trace of residual porosity can be expected for hydrogen contents below the nominal threshold values for porosity and can be detected using refined methods.[8]

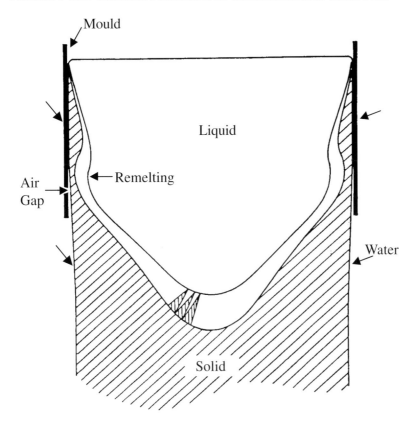

Fig. 9.5 Schematic section through aluminium alloy ingot during basic semicontinuous casting process, showing remelting of the shell due to formation of an air-gap.

9.2.2 DISTRIBUTION IN SEMICONTINUOUSLY CAST INGOTS

Hydrogen participates with other solutes in the segregation patterns that apply to particular casting process and this imposes characteristic distributions of porosity across the sections of large ingots produced commercially. For example, an ingot cast by the basic semi continuous casting process is subject to inverse segregation by the following mechanism. The initially formed shell of the ingot partially remelts when cooling is interrupted by contraction away from the mould wall as illustrated in Figure 9.5, allowing the liquid fraction from deeper in the ingot to percolate through the surface zone forming an exudation in the air gap, restoring thermal contact and consolidating the shell. The consolidated shell again contracts and the sequence is repeated. This cyclic mechanism transports liquid enriched in solutes by selective freezing into the surface zone. The surface enrichment is ultimately compensated by deficiencies at the ingot centre.

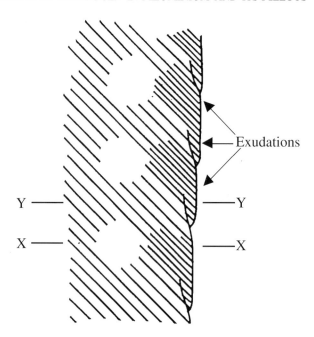

Fig. 9.6 Schematic longitudinal section through the surface of a semicontinuously cast aluminium alloy ingot showing the inverse segregation pattern associated with roots of successive exudations infiltrating the ingot shell, due to cyclic remelting. Darker zones - solute enrichment. White zones - solute depletion. XX - plane of maximum segregation. YY - plane of minimum segregation.

The frequency of the cycle is manifest by the spacing between the exudations visible on the ingot surface which is typically 2 or 3 cm and the maximum and minimum degrees of segregation occur in the planes XX and YY respectively in Figure 9.6. The quantity of porosity in the remelted surface zone is augmented by further transfer of hydrogen from solid to liquid metal fractions when short range contact between them is renewed during reversal of thermal gradients.

Figure 9.7 gives an example of the distribution of hydrogen and corresponding porosity thereby developed across the section of a 20 cm thick Duralumin type alloy (AA 2014) ingot cast by the regular semicontinuous process. The association of hydrogen with the general inverse segregation pattern is illustrated by the comparison with the distribution of copper which is also given in the figure. An unfortunate feature of Figure 9.7 is the surface porous zone about 2 cm deep that is formed under these conditions because it can become the origin of surface blemishes in subsequent wrought products. The segregation that causes this effect and other deleterious surface features can be reduced by modifying the semicontinuous casting process to delay the formation of the shell, e.g. by introducing hot top and airslip versions.

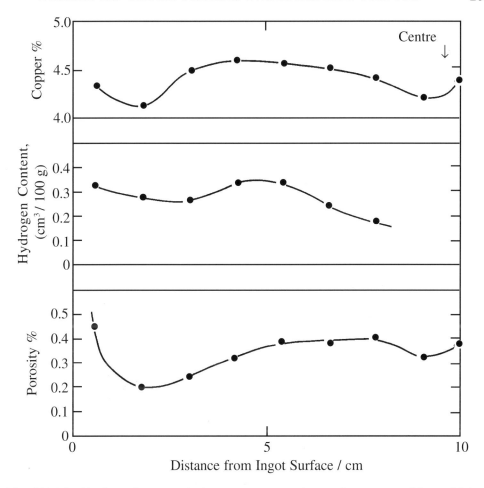

Fig. 9.7 Distribution of copper, hydrogen content and porosity across a 20 cm thick semicontinuously cast ingot of duralumin type alloy, AA 2014 cast at 8.5 cm per minute. Average hydrogen content 0.32 cm³/100 g. (Ransley and Talbot).[2]

Notwithstanding the variation of porosity across ingot sections, the relation between the *average* hydrogen content and the *average* porosity,[2] given in Figure 9.8 exhibits a threshold value and slope similar to those for the small scale AA 2014 ingots given in Figure 9.3.

The above discussion assumes that free communication between the interdendritic liquid and the external pressure is maintained throughout solidification but although it applies reasonably well to semi-continuously cast ingots in which the liquid metal remains open to the atmosphere, it is not necessarily true for castings of complex shape. Such castings may contain pockets of solidifying metal with impaired access to the liquid

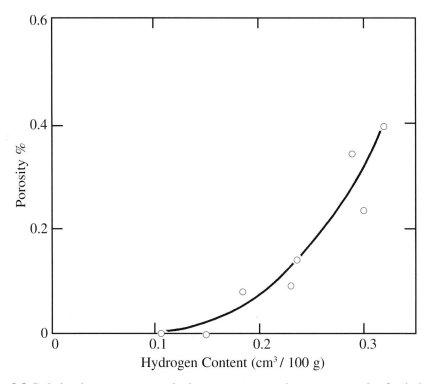

Fig. 9.8 Relation between average hydrogen content and average porosity for industrial semicontinuously cast ingots of a Duralumin type alloy to the specification AA 2014. Ingot cross-section 91 × 20 cm. (Ransley and Talbot).[2]

required to compensate for solidification contraction. In these circumstances, the interdendritic liquid is in tension due to the contraction, reducing the critical concentration of hydrogen to nucleate cavities. As Chalmers[9] and others[2, 10] observed, there is no clear distinction between "gas porosity" and "shrinkage porosity"; evolution of gas and contraction of the metal are synergistic factors generating porosity but one or the other may dominate in particular circumstances.

For application to aluminium–lithium alloys the theory of interdendritic porosity must take account of the very different solubilities of hydrogen in both the liquid and solid metal. Table 9.1 re-assembles some of the values given in Tables 6.9, 6.12 and 6.13 for aluminium–lithium alloys at relevant temperatures above and below the solidus/ liquidus range for comparison with corresponding values for pure aluminium. The solubilities of hydrogen in the liquid metal are significantly higher but the solubility ratios are much lower for the aluminium-lithium alloys than for pure aluminium, implying that although it is difficult to reduce the hydrogen content of aluminium-lithium alloys to low values they are less susceptible to porosity than pure aluminium or other aluminium alloys.

Table 9.1 Solubilities of Hydrogen in Some Aluminium-Lithium Alloys at Temperatures Relevant to Ingot Casting - Comparison with Pure Aluminium

State	Temperature		Solubility, S/(cm³/100 g) at 101 kPa (1 atm)				
	°C	T/K	Pure Al	Al - 1%Li	Al - 2%Li	Al - 3%Li	AA 2090
Solid	600	873	0.03	0.84	1.01	1.63	1.29
Liquid	700	973	0.88	2.53	2.78	3.55	5.24
Ratio, [S (Liquid)]/[S (Solid)]			30	3.0	2.8	2.2	4.1

9.2.3 Persistence Through Thermal and Mechanical Processing

Interdendritic porosity in ingots cast for fabrication does not weld up completely in the course of hot working operations but is flattened into planar discontinuities.[11–13] Hot-working sequences can be modified to promote more complete welding of the porosity. Forging before rolling[13] and two stage forging with an intermediate heat treatment to dissolve hydrogen compressed in the flattened pores[11] are beneficial but probably impracticable and uneconomic.

9.2.4 Porosity in Near Net Shape Castings

Interdendritic porosity forms in castings by the same principles as for semicontinuously cast ingots but with some important differences. Casting alloys are formulated for fluidity at low temperatures, minimum solidification contraction and ability to accept detailed imprints from the mould. The most important commercial casting alloys[14] contain silicon as a major component and solidify with a large eutectic fraction in which hydrogen bubbles can nucleate, unimpeded by a continuous dendritic network and consequently the pores tend to be spherical.[15] In contrast, high strength aluminium-magnesium alloys are very sensitive to hydrogen porosity[15–18] because of their long freezing ranges.

In sand casting, a further contribution to porosity is "mould reaction", a term that describes hydrogen absorption by the metal while it is in the mould, promoting the formation of zones of severe porosity close to the surface of castings.[19–22] The source of hydrogen is reaction between the solidifying metal and steam from residual moisture in the mould vaporised at the sand face. Pure aluminium and most alloys are protected from attack by the oxide film formed on the metal and serious damage occurs only in aluminium-magnesium alloys because the reaction products do not protect the metal from continuing attack. The attack increases in severity as solidification proceeds because the reactivity of the liquid fraction increases as it becomes enriched in magnesium by selective freezing and is most intense just above the solidus temperature.[19] It can be controlled by the methods described in Chapter 10.

9.3 SECONDARY POROSITY

9.3.1 NUCLEATION OF MICROPORES

9.3.1.1 Disproportionation of Hydrogen Solutions

In semicontinuously cast aluminium or an aluminium alloy, a hydrogen content < 0.15 cm³/100 g is usually insufficient to generate macroscopic porosity by selective freezing and is initially retained in solution in the solidifying metal. Nevertheless, the activity of even such a low hydrogen content in solid solution is very high and the theoretical arguments in Chapter 4, Section 4.2 predict that such a solution is unstable, tending to disproportionate spontaneously, yielding a heterogeneous system in which a disseminated internal gas phase accommodated in micropores is in equilibrium with a reduced solute concentration. The relevance of this theory to features of manufactured products is now addressed. Following standard practice for phases nucleated at temperatures below the solidus, it is convenient to describe a system of micropores formed in this way as *secondary porosity* to distinguish it from interdendritic porosity, i.e. *primary porosity*, that is nucleated in the liquid fraction during solidification.

Spherical micropores with typical radii of 1 to 2 μm widely disseminated throughout the structure are a common feature of semicontinuously cast ingots, illustrated for a commercial grade of aluminium in Figure 9.11. The existence of these pores and their persistence to wrought products was known empirically in advance of the theoretical treatment.[5] They were associated with the low mobility of the solute described in Chapter 7, Section 7.3.2 and were recognised as the nuclei on which large spherical pores can develop in solid metal exposed to massive hydrogen absorption in polluted atmospheres as described in Chapter 8, Section 8.4.2.4.

9.3.1.2 Stability of Micropores

Assuming that the surface tension of solid aluminium, γ, is 1.22 N m⁻¹, pores in solid aluminium with radii as small as 2 μm require an internal hydrogen pressure of about 25 atm. (2.5 MPa) to stabilise them against collapse under surface forces. Sufficient pressure is available from the rise in the activity of dissolved hydrogen due to the reduction in solubility as the result of solidification and subsequent cooling. As an example, application of Sieverts' equation, eqn 4.6, shows that a hydrogen content of 0.10 cm³/100 g in pure aluminium delivers a pressure of $(0.10/0.02)^2 = 25$ atm. when the temperature has fallen to 575°C (848 K), where the solubility in equilibrium with 1 atm. (101 kPa) given by eqn 6.7 is 0.02 cm³/100 g.

The nucleation of micropores at a stable size is probably assisted by a transient excess of vacancies quenched into the metal as it solidifies and cools from the solidus temperature.[23-27] There are two sources of excess vacancies in newly solidified metal, those introduced by the freezing process itself[23] and the cumulative excess from the rapidly diminishing equilibrium thermal vacancy population, as the metal cools, from an initial mole fraction of 9×10^{-4} at the melting point.[24, 25]

The balance of forces is expressed in Barnes and Mazey's equation:[27]

$$\frac{2\gamma}{r} = p + \frac{RT}{V_m} \ln \frac{a}{a_o} \tag{9.1}$$

where: r is the pore radius.

   γ is the surface tension of solid aluminium.

   p is the hydrogen pressure in the pore.

   V_m is the molar volume of aluminium.

   a and a_o are the transient and equilibrium activities of vacancies.

For example, application of eqn 9.1 shows that if $\gamma = 1.22$ N m^{-1} and $T = 900$ K, an excess vacancy fraction of 0.03 can stabilise a pore with a radius as small as 0.1 μm.

The potential excess vacancy concentration is ample to generate the volume of micropores usually present in cast metal i.e. $< 0.01\%$ and a transient supersaturation sufficient to assist nucleation of the pores is feasible, despite competition from other sinks.

9.3.2 SECONDARY PORE SYSTEMS IN 99.2% ALUMINIUM (AA 1100)

The development of pore systems in 99.2% aluminium is typical of corresponding developments in commercial grades of aluminium and most common alloys. According to the application, production routes may or may not call for ingots to be *homogenised* before hot working, i.e. heating them for an extended period at high temperatures to disperse segregates. Homogenisation allows readjustment of the distribution of hydrogen between the solute and the gas phase in the pores. The pores are not annihilated by the subsequent hot and cold working sequences applied in normal production routes and the system evolves according to the thermal and mechanical treatments applied to the metal.

9.3.2.1 Effects of Thermal and Mechanical Treatments

The following case history[5] illustrates representative responses of secondary porosity to thermal and mechanical treatments, using information acquired as part of a comprehensive project to evaluate the effects of homogenisation on the properties of plate and sheet.

A semicontinuously cast ingot of 99.2% pure aluminium (AA 1100) with a rectangular section, 20×76 cm, in which no interdendritic porosity could be detected was selected for evaluation. Two identical rolling blocks, each 91 cm long were cut from the ingot. One was homogenised by heating it for 12 hours at 580°C and cooled in air and the other was left untreated. After surface machining, both blocks were preheated to 510°C and hot-rolled to 9.5 mm plate. Four sets of samples were available as follows:

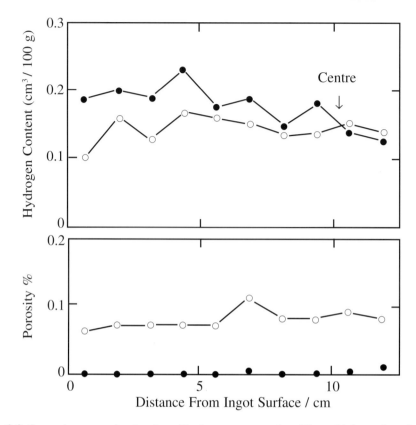

Fig. 9.9 Secondary porosity developed by heat-treatment in a 20 cm thick semicontinuously cast ingot of 99.2% pure aluminium (AA 1100).
● As cast ingot   ○   Heated for 12 hours at 580°C.

1. A 2.5 cm thick sample taken across the section of the untreated ingot.
2. A similar sample taken across the section of the heat-treated ingot.
3. Plate rolled from the untreated ingot.
4. Plate rolled from the heat-treated ingot.

Examination of Ingot Sections

Hydrogen content and porosity surveys were made across the two ingot sections with 1.25 cm sampling intervals. The hydrogen contents were measured by vacuum extraction and porosity was calculated from the densities of samples before and after consolidation as described in Section 9.1. The results are plotted in Figure 9.9. Most of the hydrogen was retained in the metal throughout the heat-treatment applied to the ingot, although some was lost by diffusion to the surface. The porosity in the cast metal was below the sensitivity of the method used to measure it, $\pm 0.01\%$, but the heat-treatment developed $0.06 - 0.11\%$ porosity throughout the ingot section.

Examination of Hot-Rolled Plates - Characteristics Inherited from the Ingot

The porosities in the hot-rolled plates derived from the untreated and heat-treated pieces of ingot were measured for comparison. Neither plate had measurable porosity, so that hot-rolling had apparently closed the porosity generated in the ingot heat-treatment.

In an experiment to discover whether hot-rolling had mechanically closed or annihilated the porosity developed in the homogenised ingot, samples of both plates were cleaned by etching, sealed with an anodic barrier film to retain hydrogen and annealed at 580 and at 500°C for progressively longer periods.

The evolution of porosity is plotted in Figure 9.10. Porosity developed in both plates but at very different rates. In plate derived from the heat-treated ingot, it rapidly approached a value of 0.1% at both 500 and 580°C, suggesting that the process was primarily the recovery of the expanded pore system inherited from the ingot that had been closed by hot rolling. Porosity in plate rolled from the as-cast ingot grew more slowly at 580°C and hardly at all at 500°C, suggesting that process was expansion *ab initio* of the pore nuclei that had persisted from the ingot to the plate.

An experiment was conducted to confirm the function of hydrogen in stabilising the porosity, as follows. Some of the samples were heated in high vacuum at 580°C to extract hydrogen. The removal of hydrogen was monitored and was complete in 2 hours, leaving the porosity unsupported. The heating was continued for a further 22 hours to allow the pores to collapse by surface tension. The porosities were redetermined and all were found to be lower, as indicated on the figure.

Micrographic Observations

Figures 9.11 and 9.12 are typical micrographs of innumerable spherical pores present in the ingot before and after homogenisation, showing that the pores expanded from ≈ 1 μm to ≈ 10 μm diameter during the heat treatment. The pores were compressed by hot rolling but persisted in the hot-rolled plate in the flattened form illustrated in Figure 9.13. On annealing the plate, the pores expanded to recover their spherical shape, as in Figure 9.14.

For selected samples, the porosity was calculated from the numbers and sizes of pores visible in microsections using standard methods,[6] allowing for the effect of random sectioning on the apparent pore size. The calculated values, given in Table 9.2, follow the same sequence as corresponding determinations from density measurements but overestimate the porosity by a factor of about 2, no doubt due to enlargement of pores in electropolishing the microsections. As the total volume of porosity increased the pores become both larger and fewer. Smaller pores are unstable with respect to larger ones because local equilibria between the surface-tension and hydrogen pressure are continually displaced by diffusion of hydrogen between them. Thus, the porosity progressively concentrates into fewer but larger pores. As the mean pore radius increases, the surface tension forces relax and the balance with gas pressure is displaced allowing the total volume of porosity to expand.

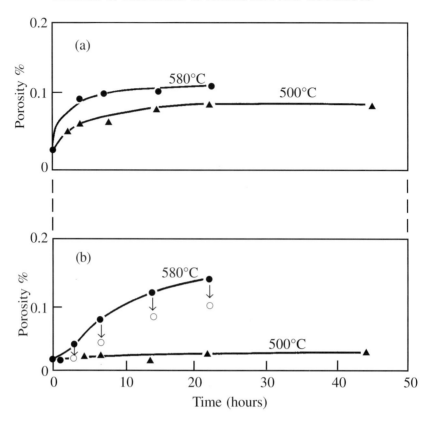

Fig. 9.10 Porosity development as functions of time in laboratory anneals at 500 and at 580°C in 9.5 mm plate, hot-rolled from ingot of 99.2% pure aluminium. (a) Plate rolled from ingot heated for 12 hours at 580°C. Hydrogen content of plate constant during anneal at 0.11 ± 0.03 cm³/100 g and (b) Plate rolled from ingot as cast. Hydrogen content of plate constant during anneal at 0.16 ± 0.03 cm³/100 g. Arrows to the open points show a decline in the porosity observed in a subsequent experiment in which the samples were re-annealed at 580°C in vacuum for 24 hours, after extracting the hydrogen.

Table 9.2 Evolution of Pore System at 500°C in 9.5 mm AA 1100 Aluminium Plate Hot Rolled from Homogenised Semicontinuously Cast Ingot (Talbot and Granger)[5]

Time at Temperature t/hours	Mean Observed Pore Radius* r/μm	Pore Count in Microsection n/cm²	Porosity % By Microscopy	Porosity % By Density
4	2.5	6.2×10^3	0.12	0.06
8	3.3	5.4×10^3	0.15	0.07
48	4.3	4.0×10^3	0.22	0.09

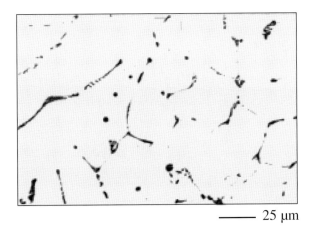

——— 25 μm

Fig. 9.11 Secondary pores in 99.2% pure aluminium (AA 1100). Electropolished section.

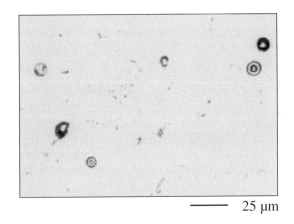

——— 25 μm

Fig. 9.12 As in Figure 9.11, after homogenisation by heating for 12 hours at 580°C.

9.3.2.2 Correlation with Hydrogen Content

The results just given in Section 9.3.2.1 referred to comparative tests on metal produced from the *same* ingot in which the hydrogen content remained virtually constant. The observations were extended to compare the incidence of secondary porosity in plates rolled from *different* ingots spanning a range of hydrogen contents, subject to the proviso

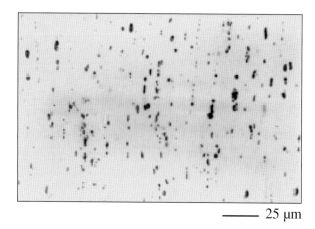

——— 25 μm

Fig. 9.13 Secondary pores in 99.2% pure aluminium (AA 1100). 9.5 mm plate hot-rolled from homogenised ingot. Electropolished section.

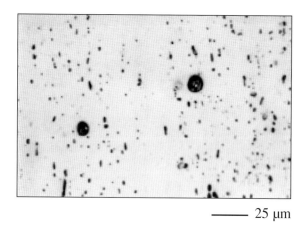

——— 25 μm

Fig. 9.14 As in Figure 9.13, annealing for 12 hours at 580°C.

that they were free from interdendritic porosity. The plates were produced as follows. Ten semicontinuously cast ingots of 99.2% aluminium (AA 1100) with 20 × 76 cm cross-section were homogenised by heating them for 12 hours at 580°C and hot-rolled to 9.5 mm plates. The plates were annealed in an electric furnace with an atmosphere of

Table 9.3 Hydrogen Contents of 9.5 cm Plates Hot-Rolled from Semicontinuously Cast 99.2% Aluminium Ingots (AA 1100) (Talbot and Granger)[5]

Ientifying Mark	Hydrogen Content (cm³/100 g)			
	Cast Ingot	Ingot Heated 12 hours at 580°C	Rolled Plate	Annealed Plate Heated to 580°C
1	Not available		0.03	0.09
2	Not available		0.04	0.12
3	0.08	0.07	0.06	0.13
4	0.07	0.06	0.06	0.14
5	0.08	0.06	0.05	0.14
6	0.18	0.11	0.15	0.23
7	0.19	0.18	0.18	0.23
8	0.19	0.16	0.17	0.28
9	0.19	0.15	0.13	0.28
10	0.18	0.13	0.15	0.29

ambient air, taking 8 hours to reach a temperature of 500°C, and air-cooled. Plates produced from all of the ingots and representative samples of their precursors at all stages of production were reserved for the measurements.

Table 9.3 gives the hydrogen contents of the metal at relevant stages of production. As expected for plates annealed under regular industrial conditions, the hydrogen contents were augmented by absorption from the rolled surface source described in Chapter 8, Section 8.5. Figure 9.15 gives corresponding values for the volumes of porosity before and after annealing, showing that they expanded to values that depended on the final hydrogen content. The role of hydrogen in stabilising the pores was again confirmed by an experiment in which hydrogen was removed from samples of the plates by heating them in high vacuum, whence the porosity collapsed, as indicated in the figure.

9.3.3 Secondary Pore Systems in Very Pure (99.99%) Aluminium

Hydrogen porosity in very pure aluminium is fundamentally different from porosity in aluminium of lower purity and alloys in two respects:
1. 99.99% aluminium has virtually no freezing range, so that there is no counterpart to interdendritic porosity in commercial grades and alloys.

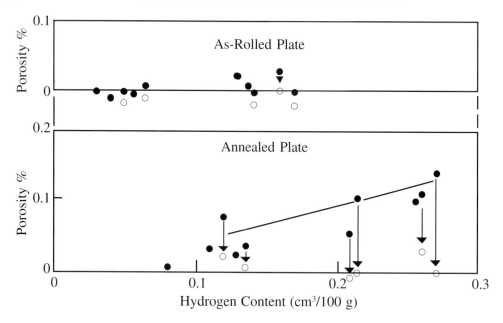

Fig. 9.15 Relation between hydrogen content and volume of latent secondary porosity developed by the heat-treatment of 99.2% aluminium (AA 1100) rolled plates. Precursor ingots heated for 12 hours at 580°C before rolling. Annealing cycle: 20°C to 580°C in 8 hours. Hydrogen absorbed during annealing displaces the experimental points to higher hydrogen contents.
- ●   Samples from plates (averages of 4 results).
- ○   Samples reheated in high vacuum for 8 hours at 540°C.

2.   When the metal is heated at high temperatures for prolonged periods, the porosity is reorganised into grain boundary networks.

9.3.3.1 Effect of Homogenising Semicontinuously Cast Ingots

As expected from the discussion in Section 9.2.1, the porosity in ingots of very pure aluminium in the as cast condition is very small, as illustrated in Figure 9.3. All of the porosity is spherical and except *in extremis* it is nucleated in the solid and can be treated as secondary porosity. An example is illustrated in Figure 9.16 which is a micrograph of a typical field in an ingot of 99.99% aluminium semicontinuously cast from liquid metal with a hydrogen content of 0.16 ± 0.02 cm³/100 g, showing innumerable pores ~1 μm in diameter randomly distributed throughout the grains.

When an ingot is heated for a prolonged period, the porosity it contains is reorganised as illustrated in the sequence of micrographs for an ingot maintained at 550 ± 2°C reproduced in Figures 9.16 to 9.18.[5] The initial pores, 1 μm in diameter, random distributed

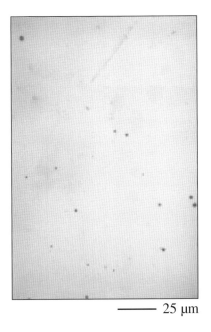

——— 25 μm

Fig. 9.16 Secondary pores in commercially produced ingot of 99.99% pure aluminium as cast. Electropolished microsection. (Talbot and Granger).[5]

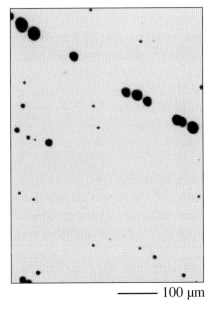

——— 100 μm

Fig. 9.17 As Figure 9.16 after heating for 4 hours at 550°C (Talbot and Granger).[5]

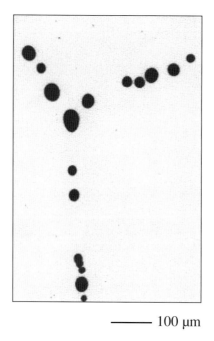

———— 100 μm

Fig. 9.18 As Figure 9.16 after heating for 16 hours at 550°C (Talbot and Granger).[5]

within the grains were progressively replaced by a grain boundary network of expanding pores that had attained diameters of ~20 μm diameter after 16 hours.

Such grain boundary pore networks in very pure aluminium can be viewed on a macroscopic scale by chemically brightening and anodising finely machined sections of ingots. This procedure gives high contrast between brightened grains and selectively etched pores at the boundaries, as in the example given in Figure 9.19.

9.3.3.2 Wrought Forms

Wrought forms produced from untreated cast ingots inherit the randomly distributed pores that persist through fabrication. These provide nuclei that can initiate the spontaneous breakdown of supersaturated solutions considered theoretically in Chapter 4, Section 4.2.

An illustration is the response of a 20 mm diameter rod, with a hydrogen content of 0.40 ± 0.02 cm³/100 g, extruded from an untreated ingot of 99.99% pure aluminium.[5] The initial volume of porosity was less than the sensitivity of measurement by the consolidation method but random pores similar to those illustrated in Figure 9.16 could be resolved in microsections. Samples were annealed at temperatures in the range 400 to 550°C, retaining the prevailing hydrogen content by sealing the cleaned surfaces with an

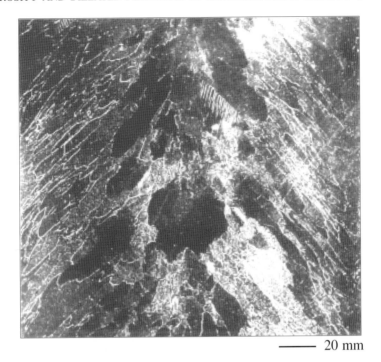

———— 20 mm

Fig. 9.19 Section of 18 cm thick ingot of 99.99% pure aluminium heated at 550°C for 96 hours. Chemically brightened and anodised to show grain boundary network of secondary

anodic film. Figure 9.20 traces the evolution of porosity in the metal. After 48 hours the volume of porosity had risen to 0.07, 0,20, 0.26 and 0.39% at 400, 450, 500 and 550°C respectively. The sensitivity to temperature is due to the combined effects of the rising diffusivity of hydrogen and the diminishing resistance of the metal to plastic flow. The initial random pore nuclei were progressively replaced by a network of pores expanded selectively at the sites of the original grain boundaries, illustrated in Figure 9.21 for a sample heated for 16 hours at 550°C. Peripheral cold work can induce recrystallisation in the surface zones of extrusions, so that the pore network does not always coincide with the final grain boundaries.

The measured volume of porosity in this sample can be correlated with the pores visible in the micrograph. The volume of porosity corresponding to the micrograph in Figure 9.21 read from Figure 9.20, was 0.23% and if most of the hydrogen content had transferred into the this volume, the gas pressure would be 1.4 MPa. This is in equilibrium with the pressure due to surface tension in pores with a radius of 1.7 µm, assuming a value of 1.22 N m^{-1} for the surface tension of aluminium, in reasonably good agreement with the radii of the pores visible in the micrograph.

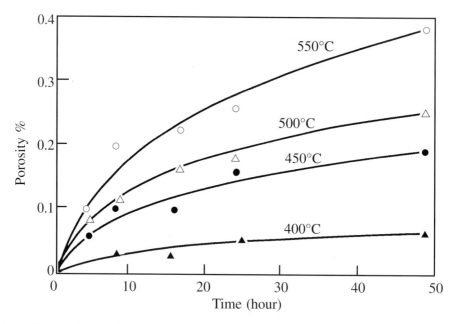

Fig. 9.20 Porosity developed in 20 mm diameter extruded rod of 99.99% pure aluminium during experimental anneal. Hydrogen content 0.40 cm³/100 g maintained constant by sealing samples with a barrier anodic film.

9.3.4 PREFERRED PORE SITES

Some explanation is needed for the difference between the evolution of pore systems in very pure aluminium (99.99%) and in AA 1100 grade aluminium (99.2%). It must be sought in factors that determine preferential pore sites.

The facility with which the total volume of porosity increases throughout very thick ingot sections, evident in Figures 9.13 and 9.20, implies that the extra void required is generated within the structure and not conducted in from the free surface. Analogy with other problems[28] suggests that this is accomplished by the generation of vacancies at local sources.

Pores close to operating vacancy sources can expand to relieve excess internal pressure by absorbing vacancies as they are generated but pores that are remote from vacancy sources can acquire vacancies only from the normal concentration in the metal and their further expansion is delayed while they are replenished by diffusion from the nearest effective sources. It is the difference in the availability of vacancy sources in very pure aluminium and AA 1100 grade aluminium that is the determining factor.

From observations such as those illustrated in Figures 9.16 to 9.18 and Figure 9.21 it is known that pores at the grain boundaries in very pure aluminium expand much more rapidly than those in the grain interiors and it follows that vacancies are more readily

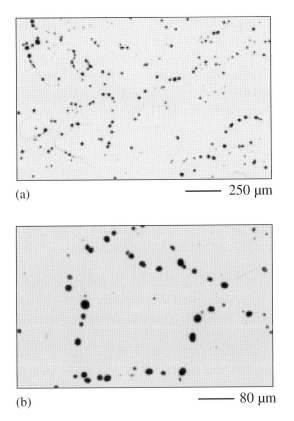

(a)   —— 250 μm

(b)   —— 80 μm

Fig. 9.21 Pore network nucleated on the original grain boundaries of extruded 99.99% pure aluminium rod heated for 16 hours at 550°C. Electropolished transverse microsection (a) General view and (b) Detail.

generated at the grain boundaries than elsewhere. A few provisional preferred pore sites in the grain interiors are apparent in Figure 9.17, which perhaps indicates minor vacancy sources at subgrain boundaries or sub microscopic particles of second phase. As the grain-boundary pores expand, the smaller pores in the grain interiors become unstable with respect to them. They therefore diminish, and the hydrogen they contain is transferred by diffusion to the grain-boundary pores. Estimates using intrinsic diffusion coefficients calculated from eqn 7.25, show that the diffusivity of hydrogen is rapid enough to allow these developments even in coarse-grained material; e.g. at 550°C, sinks separated by 1 cm could acquire almost the whole of the hydrogen content of the intervening metal in about an hour. Pores at the grain boundaries are more stable than those within the grains, since the energy of the displaced boundary is saved. Nevertheless, when the expansion of porosity has reached a point where easy supply of new vacancies is less important, the advantage of a grain-boundary site is diminished. This is shown by the persistence of some pores in the grain interiors after very prolonged heating and also by the absence of new pore development on boundaries that have moved away from established networks of pores in recrystallised metal evident in Figure 9.21.

Grain boundaries do not offer preferred sites for pore development in AA 1100 grade aluminium. This implies that pores in the grain interiors also have easy access to supplies of vacancies. The most probable source is the widely disseminated area of interphase boundary associated with the large fraction of dispersed degenerate eutectic.

9.4 REFERENCES

1. C.E. Ransley and H. Neufeld, *J. Inst. Metals*, **74**, 1948, 599.
2. C.E. Ransley and D.E.J. Talbot, *Z. Metallkunde*, **46**, 1955, 328.
3. M.F. Jordan, G.D. Denyer and A.N. Turner, *J. Inst. Metals*, **91**, 1962–63, 48.
4. B.R. Deoras and V. Kondic, *Found. Trade J.*, **100**, 1956, 361.
5. D.E.J. Talbot and D.A. Granger, *J. Inst. Metals*, **92**, 1963–64, 290.
6. K.J. Kurzydlowski and B. Ralph, *Quantitative Description of the Microstructure of Materials*, Boca Raton, Fa, CRC Press, 1995.
7. G.J. Metcalfe, *J. Inst. Metals*, **71**, 1945, 618.
8. Q.T. Fang, P.N. Anyalebechi and D.A. Granger, *Light Metals*, 1988, 477.
9. B. Chalmers, *Principles of Solidification*, John & Wiley Sons, London, 1964.
10. P.E. Doherty and B. Chalmers, *Trans. Met. Soc.*, AIME, **224**, 1962, 1124.
11. O. Kubaschewski, A. Cibula and D.C. Moore, *Gases and Metals*, (Iliffe), London, 1970.
12. J.H. O'Dette, *Trans AIME*, **208**, 1957, 924.
13. A.N. Turner and A.J. Bryant, *J. Inst. Metals*, **95**, 1967, 353.
14. John E. Hatch, ed., *Aluminium, Properties and Physical Metallurgy*, American Society for Metals, Metals Park, Ohio, 1984.
15. R.W. Ruddle and A. Cibula, *"Metallurgical Aspects of the Control of Quality in Non-Ferrous Castings"*, The Institute of Metals, London, **5**, 1957.
16. R.K. Owens, et al., *Trans. AFS.*, **65**, 1957, 424.
17. R. Jay and A. Cibula, *Foundry Trade J.*, **101**, 1956, 131 and 407.
18. W.R. Opie and N.J, Grant, *Foundry*, **78**, 1950, 104.
19. A.J. Swain, *J. Inst. Metals*, **80**, 1951–52, 125.
20. R.W. Ruddle, *Found. Trade J.*, **94**, 1953, 145.
21. M. Whittaker, *ibid.*, **95**,1953, 195.
22. M. Whittaker, *J. Inst. Metals*, **82**, 1953–4, 107.
23. P.E. Doherty and B. Chalmers, *Trans AIME*, **224**, 1962, 1124.
24. F.J. Bradshaw and S. Pearson, *Phil. Mag.*, **2**, 1957, 570.
25. R.O. Simmons and R.W. Balluffi, *Phys. Rev.*, **117**, 1960, 52.
26. F. Seitz, *Acta Met.*, **1**, 1953, 335.
27. R.S. Barnes and D.J. Mazey, *Acta Met.*, **6**, 1958, 1.
28. R.S. Barnes, *Nuclear Metallurgy*, **6**, 1959, 23.

10. Control of Hydrogen and its Effects in Manufactured Products

Manufactured aluminium products are usually destined for applications that justify their costs by exploiting their outstanding properties and aesthetically pleasing appearance. To that extent they can all be considered as critical or semicritical products, sensitive to deleterious artifacts that can compromise their integrity.

The accumulated experience and information reviewed in earlier chapters is now regularly applied to eliminate artifacts related to hydrogen. Production routes and processes are selected to match particular applications at acceptable costs. For example, more expensive processing is required for heat-treatable alloys used in aerospace applications than would be justified for half hard pure metal sheet used in domestic consumer durables. With appropriate choice of practices, problems due to hydrogen should be uncommon but unforeseen circumstances or complacency can initiate difficulties that must be resolved.

The origins of defects are multifactorial and effective countermeasures require:
1. An awareness of the defects that might be encountered.
2. Reliably controlled hydrogen contents of liquid metal delivered to ingots and castings.
3. Restricted opportunities for the solid metal to absorb additional hydrogen and to develop macroscopic defect sites during thermal treatments.
4. Care to avoid introducing extraneous defect sites during casting and fabrication.

Effective control of these factors requires inputs into the design and operation of production facilities, supported by monitoring and inspection arrangements, to limit the quantities of hydrogen occluded in the metal and to deny it the opportunity to generate defects.

10.1 DEFECTS ASSOCIATED WITH HYDROGEN

10.1.1 WROUGHT PRODUCTS

In wrought products, the nature of defects generated by hydrogen is determined by the aspect ratios imparted to them during fabrication.

10.1.1.1 Sheet and Extrusions
In heavily worked products such as sheet and extrusions of small sections, the characteristic defect related to hydrogen is blistering. The basic cause is inflation of internal defects

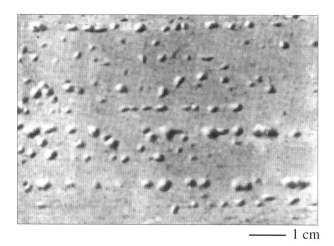

——— 1 cm

Fig. 10.1 Topography of heavily blistered sheet.

close to the surface by hydrogen pressure when the overlying metal is soft during heat treatment. Figure 10.1 gives an example of exceptionally severely blistered sheet but it is standard practice to reject any metal bearing blisters no matter how sparingly distributed or small, because freedom from blemishes that detract from aesthetic surface appearance, sensitise the metal to corrosion or jeopardise application of coatings are marketing requirements for all aluminium products.

Susceptibility to blistering is determined primarily by the nature and distribution of internal defects from any of several different origins. Sheet products and extrusions are elongated in the working direction typically by 200 times and the compressive working stresses penetrate the full section of the material as the final gauge is approached. The residues of any voids, inclusions or physical defects in the ingot persist through fabrication as long, virtually two-dimensional strings of discontinuities that can be inflated to form blisters by hydrogen already present or allowed to accumulate there by diffusion from adjacent metal.

Some typical defects providing blister sites are:
1. Residues of interdendritic porosity in ingot precursors flattened by fabrication.
2. Cavities formed by reorganisation and growth of secondary pores.
3. Non-metallic inclusions contributed from unfiltered liquid metal delivered to the ingot precursors, notably surface oxide and detritus from materials over which it flows.
4. Contamination at the interface between cladding and core in roll-bonded clad sheet
5. Coarse intermetallic particles, such as undissolved constituents of master alloys added to the liquid metal.
6. Double skin extrusion defects.

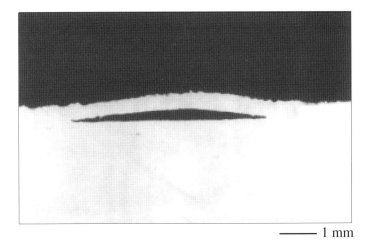

———— 1 mm

Fig. 10.2 Section through blister formed by interdendritic porosity residue.

Micrographs in Figures 10.2 to 10.6 give examples of some of these defects at blister sites.

The extent of blistering depends not only on the population of suitable defects but also on the opportunity for the overlying metal to deform into blisters. This happens when the metal is temporarily softened during heat-treatments. The most vulnerable products are therefore (i) heat-treatable alloys, which are inherently some of the stiffer alloys but which experience the highest temperature heat-treatments at the final gauge, i.e. solution-treatment preparatory to age-hardening and (ii) pure metal for deep-drawing, which is much more ductile and is fully annealed at the final gauge.

Blistering can be eliminated by current practices described later but there is no room for complacency and it is salutary to remember why they were developed. Before the Aluminium Industry acquired its present technological basis, interdendritic porosity in ingots was common because control of the hydrogen content of liquid metal was uncertain and sheet products of age-hardening alloys, e.g. AA 2024 and soft pure metal for deep-drawing could suffer variable rejections for blister averaging as much as 10% of the finished product.

Latent blister that escapes detection in products that do not receive heat-treatment at final gauge, e.g. hard temper pure metal sheet, also compromises the metal quality because it can appear in service as exfoliation corrosion. Thus the metal needs the same care in manufacture as products in which blisters would be apparent.

10.1.1.2 Foil
The microstructure of aluminium foil is often only two or three grains thick. If secondary pores expand at this stage, they force the grains apart as illustrated in the micrograph in Figure 10.7, raising them proud of the surface as illustrated in the plan view given in Figure 10.8. The effect can be produced in the special circumstance when high purity

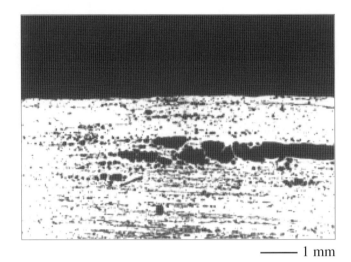

————— 1 mm

Fig. 10.3 Section through blister formed by coalescence of secondary pores in solution-treated AA 6063 alloy.

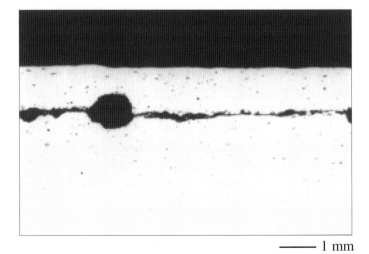

————— 1 mm

Fig. 10.4 Sections through blister at subcutaneous oxide inclusion in AA6063 alloy sheet.

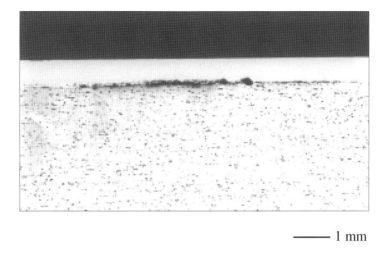

————— 1 mm

Fig. 10.5 Section through blister at fault in interface between AA 2024 alloy sheet and pure aluminium cladding.

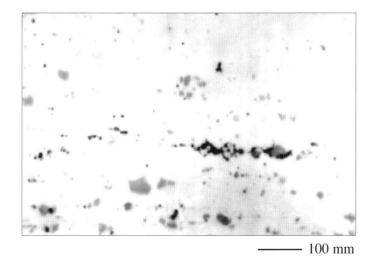

————— 100 mm

Fig. 10.6 Section through blister at internal crack through a line of coarse intermetallic particles.

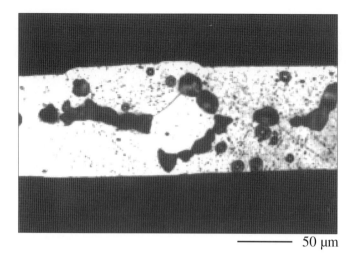

——————— 50 μm

Fig. 10.7 Section of a sample of 99.8% aluminium foil, 0.1 mm thick, taken near the edge of a coil, showing grains displaced by pores nucleated at grain boundaries by hydrogen absorbed when the coil was heated to 600°C.

foil is annealed in coil at very high temperatures *circa* 600°C to improve etching characteristics for applications in electrolytic condensers. It is associated with hydrogen absorbed during the anneal stimulated by the combined effects of a rolled surface condition and access of ambient air between loose laps of the coil. It can be controlled empirically either by chemically cleaning the foil before coiling or by providing a controlled dry atmosphere during annealing.

10.1.1.3 Thick Plate and Forgings

The mechanical properties of thick wrought products derived from defective ingots are attenuated in the short transverse direction, i.e. normal to the direction of working. The degree to which the properties are reduced depends on several factors, including the extent of deformation and the fracture toughness of the metal. A representative example is provided by Turner and Bryant's[1] measurement of the short transverse properties of 7.5 mm thick plates hot-rolled from 28 cm thick semi-continuously cast ingots of an alloy equivalent to AA 2014 solution-treated and artificially aged. The effect of successively higher values of ingot porosity was to reduce the short transverse tensile strength of the plates from the nominal value of 485 MPa to 475, 440, 430 and 420 MPa for 0.02, 0.05, 0.15 and 0.28% porosity respectively, with corresponding reduced ductility. This implies that semi continuously cast ingots produced for finished thick plates should be considered as precursors of a critical product. Cibula[2] observed similar effects for forgings in a Al-Zn-Mg-Cu alloy equivalent to AA 7075 and further found that weakness

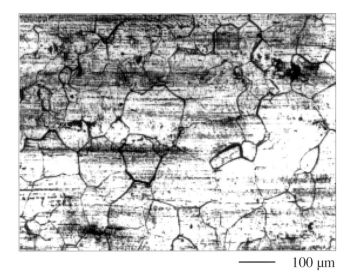

——— 100 μm

Fig. 10.8 Topography of foil shown in Figure 10.7, showing protruding grains.

induced by porosity was associated with a change in the fracture path from transgranular to intergranular, indicating internal notch sensitivity.

10.1.2 NEAR NET SHAPE CASTINGS

The foundry industry is fragmented and does not always observe the critical acceptance standards routinely fulfilled by manufacturers of wrought materials, which could improve the integrity of its products. In cast materials, the most obvious effect of defects is loss of load-bearing section and internal stress concentration due to interdendritic porosity, which is sometimes compensated by over-design.

During solidification, castings may not have continuous unrestricted access to liquid metal to compensate for contraction due to solidification and in these circumstances, shrinkage and hydrogen content co-operate in generating interdendritic porosity, so that the degree of porosity is alloy specific. For example, the properties of general purpose castings made from alloys based on the aluminium-silicon system are not unduly sensitive to moderate hydrogen contents because they have a large eutectic fraction and any porosity tends to be rounded whereas the mechanical properties of high strength aluminium-magnesium alloys are very sensitive.[3] Some of the more difficult castings to produce are those required to contain fluids under pressure in which interdendritic porosity must be avoided because it allows leaking,[4] so that the castings must be well fed and cast from metal with low hydrogen contents.

The problem of porosity generated by reaction between alloys containing magnesium and residual moisture in sand moulds, described in Chapter 9, Section 9.2.4, can be controlled by inhibitors designed to form a protective film on the metal surface at the sand face, e.g. by adding ammonium bifluoride to the sand.[2, 4] A former alternative, now prohibited by health and safety regulations was to add a trace of beryllium e.g. 0.004%, to the metal.[5]

In a former dubious practice, hydrogen was sometimes deliberately introduced into the metal to balance solidification contraction with well dispersed hydrogen porosity to offset the effects of localised shrinkage cavities in poorly fed castings.[6–8] It is now unacceptable because it is unreliable and even dispersed porosity can adversely affect the mechanical properties of a casting to an extent depending on alloy composition[3, 9] and the correct approach is to redesign the castings and their feeding arrangements.

10.1.3 Alloys Containing Lithium

As explained in Section 8.4.3.3, Chapter 8, hydrogen generated by reaction at the metal surface of aluminium-lithium alloys can precipitate lithium hydride in a surface zone as it diffuses inwards.[10, 11] The precipitate is in the form of cubic particles as illustrated in Figures 8.21 and 8.22 that have the following characteristics:
1. They are potential stress raisers sensitising the metal to early fatigue failure.
2. Setting $\Delta G^{\ominus} = 0$ in eqn 4.49 shows that they dissociate at 1140 K (867°C) under atmospheric pressure, yielding hydrogen gas.

Henry compared the fatigue lives of age-hardened Wöhler test-pieces machined from AA 2195 plate which contains 1.25 mass% of lithium with and without subsurface lithium hydride precipitation.[10] Test-pieces with lithium hydride were solution-treated in air with 0.03 atm. water vapour pressure. Test-pieces free from hydride for comparison were solution-treated in dry nitrogen. Figure 10.9 shows that the introduction of the hydride attenuated the fatigue life by nearly an order of magnitude.

Henry also assessed the effects of hydride on TIG bead-on-plate welds on samples of hot-rolled plates of AA 2195 alloy.[10] Small spherical pores, illustrated in Figure 10.10, were trapped in welds made on plates containing subsurface hydride, irrespective of whether the hydride was already present in the plate as received or introduced by heating it in humid air. No similar pores were observed in comparative welds on plates free from hydride. The volumes of the pores were consistent with the quantities of hydrogen expected from decomposition of 0.5 to 1 μm diameter lithium hydride particles at an assumed solidification temperature of about 600°C.

10.2 CONTROL AND REMOVAL OF HYDROGEN FROM THE LIQUID METAL

In view of the susceptibility of liquid metal to absorb hydrogen from furnace atmospheres and from hydrated oxides on charge materials, as described in Chapter 8, Section 8.3 it

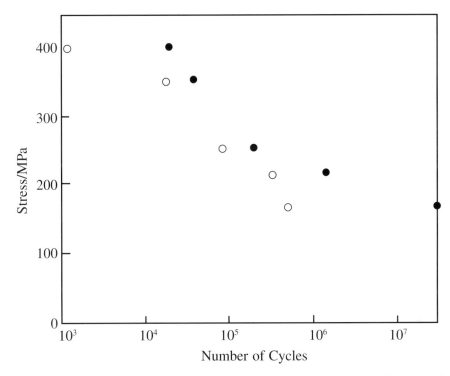

Fig. 10.9 Effect of subsurface lithium hydride precipitate on the fatigue life of an aluminium-1.25 mass% lithium alloy AA2195. Results for 3.8 mm Wöhler test-pieces solution-treated for 4 hours at 490°C, quenched and aged for 4 hours at 153°C.
- ● test-pieces solution-treated in dry nitrogen.
  - no lithium hydride precipitates detected in microsections.
- ○ test-pieces solution-treated in air with 3050 Pa (0.03 atm.) water vapour.
  - subsurface precipitates of lithium hydride observed in microsections.

is almost inevitable that industrial melts contain sufficient dissolved hydrogen to produce unacceptable porosity in ingots for fabrication or in near net shape castings. This applies both to melts prepared by remelting solid materials and to liquid metal received directly from reduction facilities.

The film formed on a clean surface of liquid pure aluminium during the first 15 or 20 minutes in humid air at typical industrial melting temperatures, e.g. 700–750°C is a matrix of η-Al_2O_3 with corundum spreading from nuclei at the metal/oxide interface. The η-Al_2O_3 dominates the oxidation mechanism during this period sustaining the transport of hydroxyl ions through its natural structural elements, thus promoting hydrogen absorption from the atmosphere as described in Section 8.3.2. When the corundum film closes over the metal/oxide interface, the process is interrupted and the hydrogen content changes only slowly thereafter.

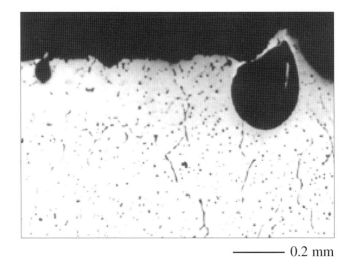

———— 0.2 mm

Fig. 10.10 Hydrogen bubbles, formed by decomposition of lithium hydride particles, trapped in the fusion zone of a TIG weld on AA 2195 plate.

The hydrogen content of the metal varies with the seasons, charge components and thermal history in the furnace. Empirically-based degassing of some kind has been in use almost since the inception of the aluminium industry but initially without clear objectives, control or evaluation. It was once erroneously assumed to be a panacea for all of the defects of the metal which could be remotely attributed to hydrogen. With more mature insight, a realistic, clear objective can now be identified as the reduction of the hydrogen content to values below threshold values of hydrogen content that will not generate interdendritic porosity as explained in Chapter 9, Section 9.2.1, typically < 0.15 cm³/100 g. Liquid metal of this quality can now be produced routinely and monitored by the methods of determining hydrogen content reviewed in Chapter 5.

10.2.1 Transport of Hydrogen Across the Liquid Metal Surface

The hydrogen contents of freshly prepared melts are typically in the range, 0.20 to 0.30 cm³/100 g, which is well above the steady state values for metal in dynamic interaction with the prevailing atmosphere, as exemplified for pure aluminium in Figures 8.4 to 8.7. Unfortunately the natural tendency of the metal to lose hydrogen is impeded by the corundum underlay in the oxide film described in Section 8.3.2.5, Chapter 8 to such an extent that it cannot be exploited as a viable means of degassing industrial melts. For the same reason, hydrogen cannot be removed from aluminium and aluminium alloys by vacuum treatments similar to those that are applied to steels.

It is possible to reduce the rate control exercised by the surface oxide on quiescent melts. The surface can fluxed, either externally by application of a molten halide mixture that dissolves the oxide or internally by oxidation of sodium in the metal, yielding the so called β-alumina which is actually the 1–3 spinel, $Na_2Al_2O_4$. Both approaches are effective but have disadvantages and are not now seriously considered. Fluxes applied to the surface deposit corundum as they dissolve oxide and there is a risk of flux inclusions in the metal. The effect of sodium is transient as it burns out of the metal, and moreover residual sodium in the metal can embrittle aluminium-magnesium alloys.[12]

10.2.2 DEGASSING BY PURGING

The principle of degassing is simply to maintain the activity of hydrogen in the gas phase below that in equilibrium with the solute and to eliminate rate-control at the gas/metal interface so that hydrogen is evolved from the metal continuously. In practice it is difficult to create conditions in which this spontaneous process occurs effectively, rapidly and reliably. All practicable systems so far developed are based on the application of an insoluble purge gas bubbled through the metal creating a very large continuously renewed internal metal surface, maintaining a hydrogen activity in the gas bubbles below that in equilibrium with the solute and eliminating rate-control at bubble surfaces, so that hydrogen is evolved continuously from the metal. The incidental turbulence helps to circulate metal so that all parts of it come into short-range contact with fresh purge gas.

10.2.2.1 Theoretical Basis
Geller's basic calculation[13] yields the theoretical maximum efficiency of degassing and identifies factors which promote effective application.

Assuming ideal behaviour for both gases and equilibrium between hydrogen in solution and in the gas phase, the volumes of hydrogen V_H, and of the purge gas, V, in the effluent are proportional to the partial pressures so that passage of an infinitesimal volume of purge gas, dV, through the metal removes a volume of hydrogen, dV_H, given by:

$$\frac{dV_H}{dV} = \frac{p_H}{(p - p_H)}$$ (10.1)

where:
   p is the total pressure on the system, assumed constant.
   p_H is the hydrogen partial pressure in the effluent.
Assuming reversibility, substituting for p_H by Sievert's relation:

$$dV = \frac{p - (C/S)^2}{(C/S)^2} \times dV_H$$ (10.2)

where:
   C is the instantaneous hydrogen content of the metal.
   S is the solubility of the gas for the pressure, p.

dV_H is proportional to the corresponding change in the hydrogen content of the metal, dC, and the mass, m, of the metal:

$$dV_H = km \cdot dC$$

Hence,

$$dV = \frac{p - (C/S)^2}{(C/S)^2} \times km \cdot dC \tag{10.3}$$

Integrating between initial and final hydrogen contents, C_I and C_F :

$$V = km \left[S^2(1/C_F - 1/C_I) + (C_I - C_F) \right] \tag{10.4}$$

where V is the volume of purge gas required to reduce the hydrogen content from an initial value, C_I, to a final value, C_F.

The most significant feature of eqn 10.4 is that the volume of purge gas, V, required to effect a given reduction of hydrogen content is proportional to the square of the solubility, S, and because S diminishes with falling temperature, degassing becomes much more effective as the temperature is reduced.

Pehlke and Bement[14] modified Geller's model to allow for the kinetics of mass transfer but, from what follows, the advantage of such refinements is illusory because it is virtually impossible to devise a single model generally applicable to real degassing systems.

10.2.2.2 Selection of Purge Gases

Degassing is a physical process and thus does not depend on the composition of the insoluble gas employed, provided it does not itself contribute hydrogen to the metal nor form films on the bubble surfaces which impede the transfer of hydrogen across the gas metal interface. The gas selected must, of course, be dry and oxygen-free. According to circumstances, it has been found convenient to use nitrogen, argon, chlorine, chlorinated hydrocarbons or mixtures.[15–19] With chlorine and chlorinated hydrocarbons such as hexachloroethane, CCl_6, the effective purging gas is aluminium chloride, $AlCl_3$. The use of hexachloroethane is convenient only for small melts because it is solid at room temperature and must be held under the metal surface to vaporise.

Chlorine is often selected or used as an additive to inert gases because it fulfils the dual role of degasser and scavenger for inclusions. This second function is important in the manufacture of high-quality products because inclusions are sources of discontinuities, blemishes, surface blisters and localised corrosion.

10.2.2.3 Degassing In Furnaces

If no special provision is made, degassing must be conducted in the melting furnace or an associated holding furnace, both of which are of reverberatory type with wide, shallow baths to promote heat transfer from the flame. Such a furnace is geometrically unsuited to the purging operation. A typical furnace holding 20 tonnes of metal has a bath area of about 12 m² and a bath depth of about 0.7 m. The purge gas is usually applied from tubes inserted manually through the side doors or, if fitted, through vertical tubes descending from the roof. In such a situation, degassing is time-consuming, inefficient and erratic and the only justification is to make the best of local circumstances when there is no other option. One or two examples suffice to illustrate the unsatisfactory nature of the operation.

Example 1 - Nitrogen Degassing Alloy AA 2014 in a Holding Furnace

Sixteen consecutive 20 tonne melts of a Duralumin-type alloy, AA 2014, were prepared over a period of several days in the same industrial equipment, comprising a melting furnace and an adjacent holding furnace to which the metal was transferred to adjust its temperature to allow detritus to settle and to degas it in preparation for casting. The purge gas was nitrogen nominally oxygen-free supplied from two cylinders, each feeding a 12.5 mm bore graphite tube inserted into the metal through a different side door. The technique of application was nominally the same for every melt. Chilled bar samples for hydrogen content determination by vacuum extraction were taken from the surface of the metal bath before and after the application of nitrogen. The results are given in Table 10.1.

The nitrogen was applied in the same manner for all of the melts. Inspection of Table 10.1 shows that the efficiency was very variable. Melts 5 to 15 were treated with comparable volumes of nitrogen, yet the reduction in hydrogen content varied from very effective for Melt 9, i.e. from 0.31 to 0.14 cm^3/100 g to ineffective for melt 11 for which the hydrogen content remained virtually unchanged at 0.26 cm^3/100 g.

Example 2 - Progress Of Nitrogen Degassing in a Holding Furnace

This example is given to show some of the characteristics of furnace degassing. The purpose of the experiment was to follow the reduction in hydrogen content as a function of time during the degassing period. The same melting equipment and method of nitrogen application was used as in Example 1, and the alloy was again AA 2014. Degassing was interrupted at 10 minute intervals to permit sampling for hydrogen content determination by vacuum extraction. The whole operation occupied 1½ hours and the metal temperature was kept reasonably constant at 670 ± 10°C. The results are given in Table 10.2 and plotted in Figure 10.11.

The process succeeded in reducing the hydrogen content to a value low enough to ensure that ingots cast from the metal would be free from gross interdendritic porosity but the with serious limitations:

1. Taking the final hydrogen content as 0.11 cm^3/100 g and the solubility, S, as 0.65 cm^3/100 g, application of eqn 10.4 demonstrates that the efficiency was only 2 or 3%.
2. The form of the curve in Figure 10.11 suggests that the reduction of hydrogen content was uneven; the apparent reversion of hydrogen content during degassing is probably a manifestation of circulation of the large mass of metal in the bath past the sampling point. With this uncertainty, the application of sufficient nitrogen to ensure that the hydrogen content was low enough, i.e. < 0.15 cm^3/100 g is very time-consuming and would entail an uneconomic reduction in output.
3. Figure 10.11 indicates that even with the high commercial grade of nitrogen used in the experiment, it is difficult to reduce the hydrogen content to a value below 0.10 cm^3/100 g.

Table 10.1 Series of Consecutive AA 2014 Alloy Melts Degassed in a Reverberatory Furnace with Nitrogen Applied through Side Doors, Metal Temperature 680 ± 10°C

Melt No.	Nitrogen Volume (m³)	Time t/minutes	Hydrogen Content (cm³/100 g)			
			Before Degassing		After Degassing	
1	4.2	30	0.28,	0.23	0.20,	0.16
2	7.9	32	0.27,	0.24	0.24,	0.21
3	9.3	39	0.34,	0.32	0.21,	0.20
4	9.3	33	0.33,	0.31	0.23,	0.21
5	11.2	66	0.28		0.20	
6	12.6	68	0.33,	0.29	0.21,	0.19
7	12.6	66	0.30,	0.26	0.22,	0.20
8	13.0	62	0.35,	0.30	0.24,	0.21
9	13.9	60	0.33,	0.29	0.14,	0.14
10	13.9	73	0.28,	0.25	0.17,	0.16
11	13.9	59	0.26,	0.26	0.25	
12	13.9	45	0.31,	0.29	0.17,	0.16
13	13.9	76	0.26,	0.24	0.16,	0.15
14	13.9	69	0.25		0.18	
15	13.9	67	0.30,	0.27	0.19,	0.17
16	27.9	80	0.28,	0.25	0.14,	0.10

Table 10.2 Progressive Reduction of the Hydrogen Content of a AA 2014 Alloy Melt During Degassing in a Reverberatory Furnace with Nitrogen Applied by Tubes through Side Doors

Time (min)	Nitrogen Volume (m³)	Temperature (°C)	Hydrogen Content (cm³/100 g)	
0	0	680	0.29,	0.25
10	3.1	680	0.19,	0.19
20	6.2	675	0.16,	0.16
30	9.3	670	0.17,	0.14
40	12.4	670	0.20	
50	15.5	670	0.19	
60	18.6	660	0.18	
70	21.7	660	0.14	
80	24.8	660	0.15	
90	27.9	660	0.11	

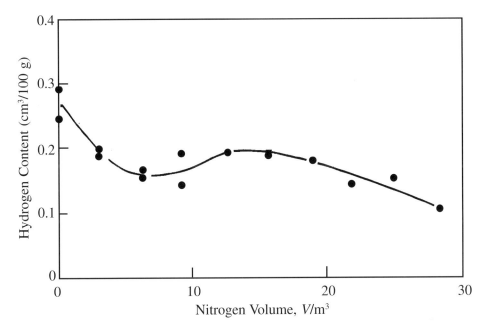

Fig. 10.11 Hydrogen content of samples taken from 20 tonnes of liquid AA 2014 alloy as a function of the volume of nitrogen applied as a purge gas. Nominally oxygen-free nitrogen applied by two 12.5 mm bore graphite tubes passed through furnace side doors and inserted into the metal.
Treatment time - 1½ hours. Metal temperature - 670 ± 10°C.

A further shortcoming of furnace degassing with nitrogen was revealed in other similar tests in which the samples taken during the degassing operation contained so many oxide inclusions that they were unsuitable for hydrogen content determinations.

This directs attention to another important aspect of purge gas degassing, i.e. the production of oxide as the bubbles break the liquid metal surface. It is for this reason that chlorine or nitrogen/chlorine mixtures are preferred as purge gases, because the bubbles carry chloride films on their surfaces and are thereby self-fluxing. At aluminium alloy melting temperatures aluminium trichloride is a gas but the more stable chloride of the common alloying element, magnesium, forms eutectic systems with other metal chlorides, yielding liquid phases that are solvents for oxides. These considerations inevitably bring together degassing and metal cleanliness as inseparable aspects of liquid metal treatment.

The reason for all of these disappointing observations is, of course, that melting and holding furnaces are designed for good thermal transfer and ease of operation and are totally incompatible with the requirements for purge gas degassing.

10.2.2.4 In-line Degassing

Most of the difficulties quoted for the obsolescent practice of degassing applied in the furnace are resolved by "in-line degassing" i.e. simultaneous degassing and filtration of the liquid metal during transfer from the furnace to the mould.[19–27] The principle is to apply a purge gas countercurrent to the metal as it passes through a suitable heated vessel. This approach has the important advantages enumerated below:

1. Countercurrent application utilises the purging gas more effectively and the filter bed improves its distribution through the metal.[23]
2. The metal does not reabsorb hydrogen from the environment before entering the mould.
3. Fume emission can be eliminated, saving the costs of fume treatment for pollution control associated with chlorine treatment in a furnace.
4. Better use is made of furnace availability.
5. If the purge gas contains chlorine it can assist filtration by back-flushing the filter bed.

An early prototype system, the Brondyke and Hess Combination Process (ALCOA 181 Process), was a Brondyke and Hess filter,[22] containing a 50 – 70 mm deep filter bed of granular material fitted with a diffuser through which argon or nitrogen is introduced to pass upward through the filter bed countercurrent to the metal flow, promoting synergy between degassing and filtration. The process was capable of producing metal of the quality required for critical applications but only if assisted by some chlorine treatment applied in the furnace. A subsequent development, the ALCOA 469 process described by Blayden and Brondyke[23] and illustrated in Figure 10.12 is more effective and can produce metal of the required quality without supplementary furnace treatment. This process employs two filter units in sequence in the metal transfer system, a primary unit filled with 2 cm diameter alumina balls acting as a roughing filter followed by a secondary unit containing a 15 – 25 cm bed of 3 – 6 mm alumina flake supported on a bed of aluminium balls. The purging gas is argon containing a 0.01 – 0.1 volume fraction of chlorine and it is applied through diffusers at the base of both units, rising through the filter beds countercurrent to the metal flow. With correct control of the chlorine addition the system does not emit fume and yet is efficient in removing hydrogen and inclusions from the metal together with sodium and calcium which are undesirable contaminants in certain alloys.

Another prototype system, the British Aluminium Company's Fumeless In-line Degassing (FILD) Process, described by Brant, Bone and Emley[19] uses a countercurrent flow of nitrogen and removes inclusions by admitting the metal through a cover of molten chloride flux floating on the surface and then passing it over flux-coated alumina balls. The general arrangement is shown in Figure 10.13. The function of the dry alumina balls on the outgoing side of the unit is to prevent flux from being carried over in the metal. Emley and Subramanian[20] suggest that the flux probably also assists degassing, quoting Kurfman's assertion that evolution of hydrogen from the metal surface is impeded by an oxide film but not by molten chlorides.[21]

An alternative principle for in-line degassing, now in vogue, is application of purge gases through rotating impellers, typically made from graphite that break the gas stream

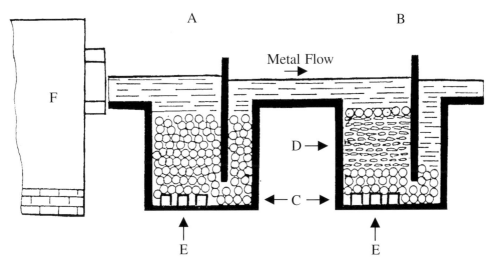

Fig. 10.12 ALCOA 469 in-line degassing process.
A. Primary Unit   B. Secondary Unit   C. Alumina Balls
D. Alumina Flake   E. Purge Gas Dispensers   F. Furnace

into fine bubbles dispersed in a deep vessel inserted in the transfer system. Using chlorine mixed with argon as the purge gas, inclusions are also removed by flotation. Representative processes are the Union carbide SNIF™ process[24] and the ALCOA 622 process[25–27] illustrated schematically in Figure 10.14, and there are other commercially available systems. Differences lie in the design of impellers, the number of stages and combinations with other treatment systems, such as deep bed filters.

All of the available in-line systems are designed to deliver large quantities of clean metal having hydrogen contents well below 0.15 cm³/100 g, to ensure that semi-continuously cast ingots of most alloys are free from interdendritic porosity. The choice between them probably depends less on perceived differences in performance than on local circumstances, product mix and layout of the equipment for which they are required.

10.2.2.5 Economical Use of Degassing Procedures

The cost of degassing in terms of the complexity of the system required, consumption of purge gas, fume treatments etc. is minimised by optimising the function expected from it. Feed back from in-line monitoring of the hydrogen content of the outgoing metal using a recording Telegas instrument can be used to control the volume of purge gas to the quantity needed.

The hydrogen content of the metal delivered to the degassing system is an obvious factor in its economical use. Mill scrap in the furnace charge must be accepted as it is because cleaning it to remove the contaminated surface is uneconomic but its contribution

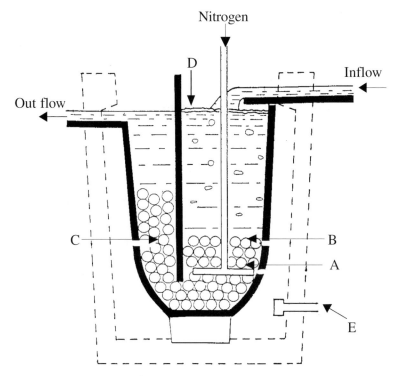

Fig. 10.13 British aluminium company's FILD in-line degassing process.
A   Purge Gas Diffuser   B   Flux Coated Alumina Balls
C   Uncoated Alumina Balls   D   Liquid Flux   E   Gas Burner

to the hydrogen content of remelted metal can be minimised by attention to fabrication to minimise scrap arisings. Low furnace temperatures keep the hydrogen solubility low but this must be balanced with the need to maintain throughput. A suitable compromise temperature is about 730°C for grades of pure metal and rather lower for more highly alloyed materials. For pure aluminium and alloys without components that oxidise selectively, hydrogen is absorbed strongly from the furnace atmosphere only when the transition alumina, η-Al_2O_3, extends to the metal/oxide interface during the short initial period after cleaning the surface as explained in Section 8.3.2.5, and it is desirable to maintain quiescent liquid surfaces to minimise reformation of η-Al_2O_3.

10.2.2.6 Multipurpose Liquid Metal Treatment Systems

An additional function of in-line degassing systems is their ability to filter entrained oxide from the liquid metal. The surface of liquid metal prepared for casting is covered with dross and unless care is exercised to prevent the liquid from enveloping its own surface during transfer between furnaces and to ingot moulds or castings, oxide fragments can become

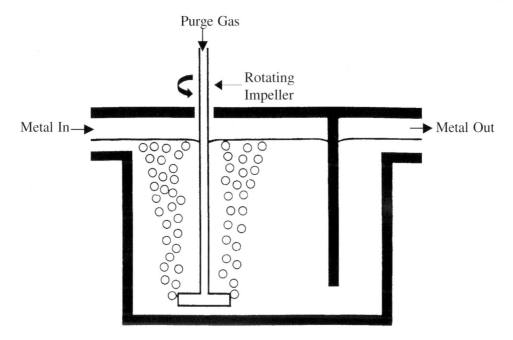

Fig. 10.14 ALCOA 622 spinning nozzle in-line degassing process. Schematic diagram.

entrained in the liquid and hence included in the structure of the subsequent solid metal. Remelt metal and in-house scrap added to the charge contributes additional inclusions. Oxide fragments are not only defects in their own right as foreign bodies but also the origin of local discontinuities formed during fabrication that can trap hydrogen. They are also likely to contain the water stabilised transition alumina, η-Al_2O_3 formed on the metal surface as described in Section 8.3.2.5, so that the discontinuities have built-in local hydrogen sources independent of the hydrogen content in the rest of the metal. Thus in-line degassing is conceived as one aspect of total in-line metal treatments which also includes filtration and metal cleaning, usually by chloride fluxes or chlorine gas.

10.2.2.7 The Special Case of Alloys Containing Lithium

In developing procedures for producing aluminium alloys with lithium as the dominant alloy component, it became apparent that the introduction of lithium radically changed the expected response of the metal to hydrogen absorbed from environmental sources. The particular observations, i.e. lower efficiency in degassing the liquid metal and yet an apparently higher tolerance for the gas in the production of defect-free material, provided the first *prima facie* evidence that the use of lithium as a component in aluminium alloys significantly enhances the solubility of hydrogen in the metal for both the liquid and solid states, a conclusion later confirmed by the results of the measurements given in Tables 6.9, 6.12

and 6.13. The uncertainty was a cause for concern because the alloys were developed for aerospace applications, where the integrity of the metal must be beyond doubt.

Attaining sufficiently low hydrogen contents to avoid interdendritic porosity and preserving them during casting is just as difficult if not more difficult than it is for conventional alloys because aluminium alloys with substantial lithium contents react more vigorously with humidity in the atmosphere and water vapour in purge gases. Information on the procedures in use is not available because they are commercially confidential.

10.3 CONTROL OF HYDROGEN AND ITS EFFECTS IN THE SOLID METAL

It is unfortunately true that although defects in unsound cast metal cannot be healed by subsequent fabrication, inappropriate or careless processing can introduce defects into initially sound metal, so that subsequent thermal and mechanical treatments must be as well chosen and carefully conducted as the production of the ingot precursors.

10.3.1 THE STATE OF OCCLUDED HYDROGEN IN THE SOLID METAL

Four occluded states of occlusion have been identified:
1. Solution of the monatomic hydrogen in the metal lattice.
2. Diatomic hydrogen rejected during solidification and trapped in interdendritic porosity.
3. Diatomic hydrogen nucleated from unstable solutions and trapped in secondary porosity.
4. Combined hydrogen chemically trapped as hydrides.

The characteristics of these forms of occlusion, discussed at length in earlier chapters, are now summarised to provide a context in which to discus methods of ameliorating deleterious effects of hydrogen on the quality of manufactured products.

10.3.1.1 Solution
When the enthalpy of dissociation of the diatomic gas is discounted from the enthalpy of solution, it is found that the dissolution of atomic hydrogen in all metals, including aluminium, is a strongly exothermic process, establishing the general principle that the hydrogen species are bound in the metal lattice by strong forces. The most useful approach is the recognition described in Section 4.1.2.1, Chapter 4, that the problem can be treated as bonding between the solute atom and the prevailing interatomic electron density in the host. In return for surrendering the aesthetic satisfaction of definitive atomistic descriptions, the embedded atom concept provides access to several matters of practical importance, notably the accommodation of hydrogen in interstitial sites, which in general are too large to maintain multiple bond contacts, hydrogen trapping at lattice defects and decohesion phenomenon where appropriate.

10.3.1.2 Interdendritic Porosity

Interdendritic porosity develops from hydrogen rejected from solution during casting as described in Section 9.2.1, Chapter 9. The quantity and distribution of the porosity is determined by the rate of heat abstraction, the selective freezing characteristics of the alloy and the segregation patterns characteristic of the casting process. It is the origin of discontinuities that persist through the very severe deformation by hot and cold working entailed in the manufacture of plate and sheet products, forgings, extrusions, and wire.[15, 28] The hydrogen in the porosity is pinched off into pockets of compressed gas parallel to the direction of working and in unidirectional heavily worked metal such as sheet or extrusions the pockets are incipient blisters that can inflate during any subsequent annealing or solution treatment of the metal. An adverse feature of interdendritic porosity in semicontinuously cast ingots is that it is concentrated near the surface as explained in Section 9.2.2, Chapter 9, so that the residues persisting to finished products are effective precursors of blisters because the overlying metal is thin.[29] Unless they become actual blisters, it is not easy to locate the residues in heavily worked extruded or rolled products.

10.3.1.3 Secondary Porosity and Unstable Solutions

Extrapolation of the critically assessed values given in Table 6.4, shows that at characteristic commercial annealing temperatures, e.g. 340 to 400°C, the solubility of hydrogen in solid pure aluminium is <0.005 cm^3/100 g but it is quite common for macroscopically sound metal in production to contain >0.15 cm^3/100 g of hydrogen. Application of Sieverts' relation suggests that if the gas were hypothetically in solution, it could yield a gas phase exerting a nominal pressure of 90 MPa, exceeding the proof stress at elevated temperatures and so it is doubtful whether such a solution could exist without disrupting the metal in the absence of a counterbalancing external pressure. The principle that the solubility is limited by the physical properties of the solvent is the basis on which nucleation and growth of secondary porosity is expected as considered in Section 4.2.1, Chapter 4.

Secondary porosity is present in embryonic form in most commercial products but it can escape detection in routine work because it is easily confused with polishing artifacts in mechanically polished microsections and with etch pits in electropolished sections. Secondary porosity has profound practical effects on the behaviour of hydrogen in manufactured products, notably by providing a field of traps that retards the diffusion of hydrogen as described in Section 7.3.2, Chapter 7 and by its capacity to accept absorbed hydrogen for reorganisation and coalescence into macroscopic defects[15, 30] causing extensive internal damage to the metal. The phenomenon is inextricably linked with the ability of the metal to accommodate hydrogen discussed in Chapter 8, Section 8.4.2.4.

In its original unmodified form secondary porosity appears to be innocuous but because it persists through mechanical working to final products, it has potential for initiating creep cavitation and internal fatigue cracks. The small cavities which Renon and Calvet[31, 32] cited as the origins of creep fissures can now, with hindsight, be identified as secondary porosity.

10.3.1.4 Hydrides

Lithium Hydride

The calculation given in Section 4.3.3.1, Chapter 4 shows that LiH is theoretically stable at temperatures < 833 K (560°C) in the presence of hydrogen at unit activity and lithium at the activities prevailing at the solvus in the binary aluminium-lithium system and yet from the empirical results given in Tables 6.12 and 6.13, it is beyond doubt that the hydride does not form immediately in the interior of the *bulk metal* when the gas is introduced at unit activity through its surface. However, the hydride forms with ease in laboratory experiments using the powdered alloy[33] and so the long-term stability of hydrogen solutions in aluminium-lithium alloys is uncertain. One cause for concern is the remote possibility that hydride might precipitate after years of incubation in stress-bearing aircraft components. As a precaution, it is advisable to eliminate hydrogen as completely as economically practicable.

A more immediate concern is to limit absorption of hydrogen during heat-treatments to avoid the formation of subsurface hydrides that can compromise fatigue life.

Sodium Hydride

The only other hydride that has been observed in aluminium or any of its commercial alloys is sodium hydride.[12] A minor effect of sodium, which is a normal contaminant in trace quantities, is to scavenge hydrogen in very pure (99.99%) aluminium but it is best to remove it from the metal by in-line liquid metal treatments so that recycled scrap cannot contaminate alloys with >3 mass % magnesium that are sensitive to sodium embrittlement.[12]

10.3.2 ABSORPTION OF HYDROGEN FROM ATMOSPHERIC HUMIDITY

10.3.2.1 Pure Aluminium

From the discussions in Sections 8.3.2.2 and 8.4.1, Chapter 8, it is apparent that a *clean* aluminium surface is remarkably resistant to the transfer of hydrogen from an atmospheric water vapour source to solution in the metal and that this is attributable to the impermeability of corundum at the metal/oxide interface to hydroxyl ions so that they cannot reach the low oxygen potential of the metal needed for their reduction. Thus significant hydrogen absorption can occur only when corundum is replaced by another oxide species amenable to the transport of hydroxyl ions.

10.3.2.2 Aluminium Alloys Containing Magnesium

For alloys with more than 1 mass% magnesium, the stable oxidation product on the metal surface is MgO and the predisposition of the metal to absorb hydrogen is determined by the characteristics of that oxide. In clean humid air, the absorption by the solid metal is tangible but slow, consistent with the known slow hydration of the oxide observed in other contexts. In humid air polluted by sulphur dioxide, the absorption is accelerated by some incompletely identified mechanism and the capacity to receive the gas is provided by expansion of pores nucleated by secondary porosity inherited from ingot precursors as discussed at length in Section 8.4.2.4, Chapter 8,

The products most at risk are age-hardening medium strength alloys in the AA 6000 series such as AA 6063 alloy in the form of extrusions and sections drawn from them, with lengths unsuited to solution-treatment in nitrate salt baths, so that they must be solution-treated in multipurpose electric furnaces. The origin of sulphur contamination in these furnaces is uncertain but it is probably accumulated from the decomposition of drawing lubricants carried in on the surfaces of sequential charges. It is not always easy to detect directly in the furnace atmosphere but when present it is easily identified indirectly from the odour of hydrogen sulphide on metal withdrawn and allowed to cool in air. The effect on the metal is manifest as blistering but it is usually a symptom of porosity throughout the section.

A simple, inexpensive and very effective inhibitor is described in Stroup's patents.[34] Small containers of potassium borofluoride, KBF_4 or sodium silicofluoride, $NaSiF_6$ are placed in the furnace to establish a trace of fluoride vapour in the atmosphere:

$$KBF_4 = KF + BF_3 \tag{10.5}$$

$$Na_2SiF_6 = 2NaF + SiF_4 \tag{10.6}$$

These compounds are chosen because their vapour pressures are of the right order at solution-treatment temperatures. The vapours establish very thin protective fluoride films on the metal surfaces as a light, not unattractive patina. There are unwanted side reactions that can stain the charge if these compounds are in contact with metals so that they are contained in non-metallic vessels and melted before insertion into the furnace to eliminate the risk of contaminating the metal charge with powder.

10.3.2.3 Aluminium Alloys Containing Lithium

The limited production of alloys containing lithium is insufficient to yield reliable information on the hydrogen absorption during manufacture. The information from the experimental work cited in Section 8.4.3.2 and 8.4.3.4, Chapter 8 that hydrogen absorption is *promoted* in unpolluted air and *inhibited* in air polluted by sulphur dioxide is quite unexpected. It will be interesting to follow this further in due course.

10.3.3 ABSORPTION OF HYDROGEN FROM SURFACES OF MILL PRODUCTS

The hydrogen potential of rolled surfaces reported in Section 8.5, Chapter 8 has both direct and indirect effects in the manufacture of aluminium products.

Direct Effects

The direct effects of rolled surfaces is to promote hydrogen absorption by metal in progress. In the production of sheet, the hot-rolled intermediate material is partly cold worked and is annealed in forced air circulation furnaces to soften it before forwarding it for cold-rolling. In a typical annealing operation, the metal is heated to 340 to 400°C, taking several hours to reach the temperature in a thermal cycle imposed by the thermal capacity of the furnace. Reference to Figures 8.27 and 8.28 shows that the metal can accept such a thermal cycle without acquiring a significant hydrogen content from the rolled surface source. However this assumes that the furnace temperature is uniform, that control sensors are correctly located and that there is no temperature overshoot by delay in removing the charge. There have been occasions where on investigation the temperature in parts of a furnace were found to be up to 100°C higher than were indicated by badly positioned sensors.

Indirect Effects

All aluminium mill products are destined for applications which are critical to some degree and must meet high standards of inspection. This leads to high rejection rates for blemished or scratched substandard material and when scrap comprising these rejections is added to mill scrap from end and edge trimming, off-cuts left after punching circular blanks, off-gauge material etc., the yield of metal for the less critical applications can be as low as 70% and the yield of materials for the more critical applications is even less.

In-house scrap is returned for remelting to recover metal values. A typical melting furnace charge can contain up to 30 or 40% of this material, the balance being new virgin metal. Figures 8.27 and 8.28 show that the mill scrap acquires a high hydrogen content before it melts. Cold-rolled sheet has the most potent surface source and this partly explains the notoriety of light scrap as a contributor of hydrogen to liquid metal. In practice, it is found that the uptake of hydrogen from light scrap is least if it is submerged in a heel of liquid metal, probably because if the metal is air melted there is an additional contribution from reaction of semi liquid material with furnace atmospheres These mechanisms are probably the primary sources of the contribution that solid charges make to the hydrogen content of liquid metal. Scrap recycled from service in the field is unreliable and it is best routed to the production of casting alloys where the tolerances are wider. Beverage cans are exceptional because the metal values are high, they can be segregated from other scrap and special arrangements are made for recycling.

10.3.4 Redistribution and Expansion of Secondary Porosity

In its embryonic form, secondary porosity is so fine that, although it is persistent through fabrication, it is probably without effect in standard products. The risk lies in treatments accorded the metal which allow it to expand or become reorganised into damaging configurations. These matters have been dealt with in detail in Section 8.4 Chapter 8 and Section 9.3, Chapter 9. Summarising, there are two practices which should be avoided, (i) excessively prolonged high-temperature heat-treatments at any stage of processing and (ii) heat-treatments of alloys containing magnesium in air furnaces when there is reason to suspect sulphur contamination.

In Section 9.3.3, it was shown that prolonged heat-treatment of very pure (99.99%) aluminium promotes not only expansion of the porosity but its re-alignment along grain and perhaps also sub-grain boundaries, to the extent that cohesion between the grains is compromised, as evident in Figures 9.16 to 9.19. The point was made that this appears to be a feature of single phase material and so it is expected to apply to solid solution strengthened alloys made on a very pure aluminium base. These alloys are produced for anodising quality bright metal products which would not normally be subject to lengthy heat-treatments. There is however a range of anodising quality alloys made with a 99.8% pure aluminium base which can be selected from the output of regular reduction cells, thus avoiding the expense of the refined purer product from the three-layer process. This material is not quite single phase but contains about 0.01 mass % of iron out of solution which reduces the brightness of the final anodised product. To overcome it, the ingots may be homogenised at high temperature (550°C) for 12 hours or more. Reorganised secondary porosity can appear as streaks marring the anodised finish of the final product. The countermeasure is a compromise, using empirically developed procedures.

In multi-phase materials, secondary porosity is not re-distributed to grain boundaries as it is for single-phase alloys, but it can expand in situ to the extent that it destroys the usefulness of the metal, usually manifest by severe blistering, as illustrated Figures 8.17 and 10.3. The effect is usually associated with hydrogen absorption during the heat-treatment of alloys containing magnesium in air furnaces.

10.3.5 MECHANICALLY-GENERATED DEFECTS

Assuming that defect-free ingots have been produced, they must be processed in such a way that new defects are not generated adventitiously because the metal is never completely free from hydrogen and it is vulnerable to the synergistic disruptive effect of hydrogen pressure and local mechanical failure.

Examples of internal mechanically-generated defects stabilised by hydrogen from solution are cracks initiated by oversized intermetallic particles remaining from incomplete dissolution of master alloys, areas of poor adhesion at the interface of roll-bonded clad sheet, and double-skinning and back-end defects in extrusions. Control of such defects is a matter of fabrication practices.

10.3.6 INHERITED CHARACTERISTICS

It is apparent from the foregoing, that it is possible to control the potential adverse effects of hydrogen in aluminium products only by conceiving production sequences as a whole. Some of the most informative results covered in this text are those illustrated in Figures 8.12 to 8.15, which show that the susceptibility of metal to hydrogen absorption at a late stage of processing is conditioned by characteristics which must have been imparted at the ingot stage. Thus if metal is initially of good quality, it is resistant to malpractice later in the production sequence. Failure to appreciate this principle of conditioned responses probably lies at the root of many abortive attempts to overcome specific metal quality problems.

Another example is the distribution of interdendritic porosity across the section of semi-continuously cast ingots and its relationship to the exudation pattern which controls its distribution at the surface. This conditions the response of sub-standard metal to blistering at the final gauge. It is obvious that the thicker is the metal overlying discontinuities, the more resistant it is to blistering, so that control of ingot casting parameters to reduce exudation, promoting deeper distribution of porosity gives a greater tolerance for lapses in the treatment of liquid metal before blisters appear on the final fabricated product.

Inheritance also runs countercurrent to the production sequence. The discussion in Section 10.3.3 of the current chapter identified recycled in-house scrap as an important source of hydrogen in liquid metal prepared for ingot casting. This can promote a self perpetuating cycle. The more rejections there are for inadequate quality, the greater is the proportion of scrap charged to melting furnaces and the more difficult becomes the control of hydrogen and oxide contents. The ratio of surface hydrogen sources to metal volume for flat rolled products increases with the degree of reduction through both hot and cold rolling, so that the following factors must be appreciated:
1. Recycling metal rejected for any cause can exacerbate problems related to hydrogen.
2. It is good practice to reject inadequate metal at the earliest stage.
3. More intensive liquid metal treatment is needed for the duration of rejection epidemics.

10.4 REFERENCES

1.   A.N. Turner and A.J. Bryant, *J. Inst. Metals*, **95**, 1967, 353.
2.   O. Kubaschewski, A. Cibula and D.C. Moore, *Gases and Metals*, (Iliffe), London, 1970.
3.   R. Jay and A. Cibula, *Foundry Trade J.*, **101**, 1956, 131 and 407.
4.   R.W. Ruddle and A. Cibula, *J. Inst. Metals*, **85**, 1956–7, 265.
5.   Marjorie Whittaker, *J. Inst. Metals*, **82**, 1953–4, 107.
6.   E. Scheuer, *J. Inst Metals*, **85**, 1956–7, 521.
7.   British Patent No. 685075.
8.   British Patent No. 544560.
9.   W.R. Opie and N.J. Grant, *The Foundry*, **78**, 1950, 104.
10. D.M. Henry, *"The Nature and Effects of Hydrogen in Weldalite Aerospace Alloy and Other Aluminium-Lithium Alloys"*, Ph.D. Thesis, Brunel University, 1995.
11. R.C. Dickenson, K.R. Lawless and K. Wefers, *Scripta Metallurgica, 22,* 1988, 917.
12. C.E. Ransley and D.E.J. Talbot, *J. Inst. Metals*, **88**, 1959–60, 150.
13. W. Geller, *Z. Metallkunde*, **41**, 1950, 124.

14. R.D. Pehlke and A.L. Bement, *Trans. AIME*, **224**, 1962, 1237.
15. H. Kostron, *Z. Metallkunde*, **43**, 1952, 269 and 373.
16. L.W. Eastwood, *Gas in Light Alloys*, Chapman and Hall, London, 1946.
17. L Sokol'skaya, *Gas in Light Metals*, Pergamon Press, London, 1961.
18. M. Tikkanen and E. Erkko, *Avgasning av aluminiumsmaltor medelst gasinblasnung*, IVA, **27**, 1956, 96.
19. M.V. Brant, D.C. Bone and E.F. Emley, *J. Met., March,* 1971, 48.
20. E.F. Emley and V. Subramanian, *Light Metals*, **2**, 1974, 649.
21. B. Kurfman, *Modern Castings*, **45**, 1964, 9.
22. K.J. Brondyke and P.D. Hess, *Trans. AIME*, **230**, 1964, 1553.
23. L.C. Blayden and K.J. Brondyke, *Light Metals*, **2**, 1973, 493 AIME and US Patent Nos. 3737303-3737305.
24. A.G. Szekely, US Patent Nos. 3227547, 3743263, 3870511.
25. M.J. Bruno, N. Jarret, B.L. Slaugenhaupt and R.E. Grazlano, US Patent No. 3839019.
26. R.E. Miller, L.C. Blayden, M.J. Bruno, C.E. Brooks, *Light Metals*, AIME, **2**, 1978, 491.
27. L.C. Blayden and K.J. Brondyke, *J. Metals*, **26**, 1974, 1.
28. J.H. O'Dette, *Trans AIME*, **208**, 1957, 924.
29. C.E. Ransley and D.E.J. Talbot, *Z. Metallkunde*, **46**, 1955, 328.
30. H. Chadwick, *J. Inst. Metals*, **83**, 1954–55, 513.
31. C. Renon and J. Calvet, *Mem. Sci. Rev. Met.*, **58**, 1961, 835.
32. C. Renon and J. Calvet, *Mem. Sci. Rev. Met.*, **60**, 1963, 620.
33. S. Aronson and F.J. Salzano, *Irong. Chem.*, **8**, 1969, 1541.
34. P.T. Stroup, US Patents Nos. 2092033 and 2092034.

Appendix 1. Calibration of Vacuum Extraction Equipment

The only opportunity to calibrate the McLeod gauge and the volume of the collection system of the vacuum extraction equipment described in Chapter 5 is before the equipment is finally assembled and commissioned and therefore it must be done very carefully.

Calibration of McLeod Gauge (J in Figure 5.1)

Capillary Tube Radius, r
The capillary tube in the closed limb is about 100 mm long and is purchased with a precision bore, guaranteed to be 1 ± 0.01 mm in diameter. All that is required is to confirm the diameter by calculating it from the mass and length of a thread of mercury introduced into the bore, filling it completely. The top of the capillary tube is left open to permit entry of the mercury thread. The length of the thread is measured with a travelling microscope, making the usual corrections for the menisci at either end. It is transferred to a suitable miniature container and weighed on a precision balance. After several replicate measurements, the top of the capillary tube is sealed to give a flat closed end.

Gauge Volume, V
The volume of the closed limb is about 100 cm^3. It measured by titration with air-free distilled water, using a precision (Class A) burette. The gauge is inverted and the water is run in until it just reaches the position corresponding to the position where the mercury cuts off the volume when the gauge is in use. The average of several replicate measurements is taken and the volume of the capillary tube that the water does not enter is added. The scatter between replicate results is expected to be $< \pm 0.1\%$. After calibration, the gauge is installed in the vacuum system.

Calibration Factor
The McLeod gauge reading is the difference in height between the mercury columns in the open and closed capillary tubes when the mercury is raised in the open capillary tube to be level with the top of the closed capillary tube. The calibration factor has the dimension, (length)$^{-1}$. Conventionally, the reading is quoted as h/mm so that the calibration parameters must be expressed in the same unit of length, i.e. the radius of the capillary tube is r/mm and the gauge volume is V/mm^3. Application of Boyle's law, yields for the system pressure, p, expressed in the correct SI unit, the Pascal:

$$p/\mathrm{Pa} = \frac{\pi(r/\mathrm{mm})^2}{V/\mathrm{mm}^3} \times \frac{101325}{760} \times (h/\mathrm{mm})^2 = \mathrm{K}\,(h/\mathrm{mm})^2 \qquad (\mathrm{A1.1})$$

The value of K for a particular gauge is its calibration factor and is of the order of 1×10^{-3}

Calibration of the Collection System Volume

The limits of the collection system volume, V_S are the tap, P, and the back of the diffusion pump, G, in Figure 5.1. It is calibrated by the expansion of hydrogen into it from a known auxiliary volume, V_B. The auxiliary volume is provided by a glass bulb joined by glass-blowing through an isolating tap to the collection system. For adequate sensitivity, V_B is approximately equal to the expected value of V_S.

The composite system, comprising $V_S + V_B$ is evacuated and tested for vacuum tightness. With the pump G operating and tap P closed, hydrogen is admitted to the composite system by gently stroking the palladium tube, L, with a very small hydrogen flame, taking care not to overheat it. The quantity of hydrogen required corresponds to a Mcleod gauge reading near the limit of its scale. To reduce the high concentration of hydrogen that dissolves in the palladium during this operation, the flame is removed and the tube is warmed by its small heater, until the pressure in the system just begins to fall. The pressure, p_1, is allowed to stabilise and measured several times in succession with the McLeod gauge to ensure that it is known accurately. For precise readings, it is advisable to tap the closed capillary tube very gently to ensure that the mercury meniscus is at its equilibrium position.

The tap to the auxiliary volume is closed to retain the hydrogen it contains and tap P is opened to evacuate the collection system. Tap P is closed again and the tap to the auxiliary volume is opened, allowing the hydrogen in the volume, V_B to expand into the composite volume $(V_S + V_B)$. The new pressure, p_2 is measured several times in succession as before. Application of Boyle's law gives the collection system volume:

$$V_S = \frac{p_1 - p_2}{p_2} \times V_B \qquad (\mathrm{A1.2})$$

Results of replicate measurements are averaged. With care, the scatter between the replicate results is < 2%.

Finally, the vacuum in the system is relieved, the glass bulb and its tap are sealed off and removed by glassworking, and the system is re-evacuated.

Appendix 2. Selective Oxidation of Magnesium and Lithium in Alloys

A2.1 STANDARD GIBBS FREE ENERGIES OF FORMATION

Aluminium - magnesium - oxygen - water system

$$\frac{4}{3}Al(s) + O_2(g) = \frac{2}{3}Al_2O_3(s)$$

$$\Delta G_1^{\ominus} = -1117993 - 11.1\ T \log T + 244.5\ T \text{ Joules} \tag{A2.1}$$

$$2Mg(s) + O_2(g) = 2MgO(s)$$

$$\Delta G_2^{\ominus} = -1207921 - 24.7\ T \log T + 284\ T \text{ Joules} \tag{A2.2}$$

$$MgO(s) + Al_2O_3(s) = MgAl_2O_4(s)$$

$$\Delta G_3^{\ominus} = -25000 - 2.1\ T \text{ Joules} \tag{A2.3}$$

$$\tfrac{1}{2}Mg(s) + Al(s) + O_2(g) = \tfrac{1}{2}MgAl_2O_4(s)$$

$$\Delta G_4^{\ominus} = \tfrac{3}{4}\Delta G_1^{\ominus} + \tfrac{1}{4}\Delta G_2^{\ominus} + \tfrac{1}{2}\Delta G_3^{\ominus}$$

$$= -1152974 - 14.5\ T \log T + 253.4\ T \text{ Joules} \tag{A2.4}$$

$$MgO(s) + H_2O(g) = Mg(OH)_2(s)$$

$$\Delta G_5^{\ominus} = -46024 + 100\ T \text{ Joules} \tag{A2.5}$$

$$2Mg(s) + O_2(g) + 2H_2O(g) = 2Mg(OH)_2$$

$$\Delta G_6^{\ominus} = G_2^{\ominus} + 2\Delta G_5^{\ominus}$$

$$= -1299969 - 24.7\ T \log T + 484\ T \text{ Joules} \tag{A2.6}$$

Aluminium - lithium - oxygen - water system

$$4Li(l) + O_2(g) = 2Li_2O(s)$$

$$\Delta G_7^{\ominus} = -1182172 + 85.8\ T \log T - 3.85\ T \text{ Joules} \tag{A2.7}$$

$Li_2O + Al_2O_3 = Li_2Al_2O_4$

$\Delta G_8^\ominus = -108000 + 17.4\,T$ Joules   (A2.8)

$Li(l) + Al(s) + O_2(g) = \frac{1}{2}Li_2Al_2O_4(s)$

$\Delta G_9^\ominus = \frac{3}{4}\Delta G_1^\ominus + \frac{1}{4}\Delta G_7^\ominus + \frac{1}{2}\Delta G_8^\ominus$

$\qquad = -1188000 + 13.1\,T \log T + 174\,T$ Joules   (A2.9)

$Li_2O(s) + H_2O(g) = 2LiOH(s)$

$\Delta G_{10}^\ominus = -146189 + 108.7\,T$ Joules   (A2.10)

$4Li + O_2 + 2H_2O = 4LiOH$

$\Delta G_{11}^\ominus = G_7^\ominus + 2\Delta G_{10}^\ominus$

$\qquad = -1474550 + 85.8\,T \log T + 214\,T$ Joules   (A2.11)

$\Delta G_1^\ominus$, $\Delta G_2^\ominus$, $\Delta G_5^\ominus$, $\Delta G_6^\ominus$, $\Delta G_{10}^\ominus$ are given explicitly.[1]

The following approximations are used for $\Delta G_3^\ominus$ and $\Delta G_8^\ominus$:

$\Delta G_3^\ominus = \Delta H_3^\ominus\,(298\text{ K}) - T\Delta S_3^\ominus\,(298\text{ K})$

$\qquad = \Delta H_3^\ominus\,(298\text{ K}) - T[S^\ominus\,(MgAl_2O_4,\ 298\text{ K}) - S^\ominus\,(MgO,\ 298\text{ K}) - S^\ominus\,(Al_2O_3,\ 298\text{ K})]$

$\qquad = -25000^{[1]} - T(80.5 - 27.4 - 51.0)^{[3,4]}$

$\qquad = -25000 - 2.1\,T$ Joules

$\Delta G_8^\ominus = \Delta H_8^\ominus\,(298\text{ K}) - T\Delta S_8^\ominus\,(298\text{ K})$

$\qquad = \Delta H_8^\ominus\,(298\text{ K}) - T[S^\ominus\,(Li_2Al_2O_4,\ 298\text{ K}) - S^\ominus\,(Li_2O,\ 298\text{ K}) - S^\ominus\,(Al_2O_3,\ 298\text{ K})]$

$\qquad = -108000^{[5]} - T(106.3 - 37.9 - 51.0)^{[4,6,7]}$

$\qquad = -108000 - 17.4\,T$ Joules

A2.2 Formation of Oxides on Aluminium-Magnesium Alloys

A2.2.1 Equilibrium Oxygen Activities

The following assumptions are made:

1. The oxides are pure so that $a_{Al_2O_3}$, a_{MgO} and $a_{MgAl_2O_4}$ are all unity.
2. Alloys considered lie within aluminium-rich solid solution phase fields.

A2.2.1.1 Equilibrium Constants

Applying the Van't Hoff isobar to eqn A2.1:

$$\ln K_{Al_2O_3} = -\frac{\Delta G_1^{\ominus}}{RT} - \frac{(1117993\,T^{-1} - 11.1 \log T + 244.5)}{8.314} \qquad (A2.12)$$

where: $K_{Al_2O_3} = \dfrac{1}{(a_{Al})^{\frac{4}{3}} \times a_{o_2}}$ $\qquad (A2.13)$

Applying the Van't Hoff isobar to eqn A2.2:

$$\ln K_{MgO} = -\frac{\Delta G_2^{\ominus}}{RT} - \frac{(1207921\,T^{-1} - 24.7 \log T + 284)}{8.314} \qquad (A2.14)$$

where: $K_{MgO} = \dfrac{1}{(a_{Mg})^{2} \times a_{o_2}}$ $\qquad (A2.15)$

Applying the Van't Hoff isobar to eqn A2.4:

$$\ln K_{MgAl_2O_4} = -\frac{\Delta G_4^{\ominus}}{RT} - \frac{(1152974\,T^{-1} - 14.5 \log T + 253.4)}{8.314} \qquad (A2.16)$$

where: $K_{MgAl_2O_4} = \dfrac{1}{(a_{Mg})^{\frac{1}{2}} \times (a_{Al}) \times a_{o_2}}$ $\qquad (A2.17)$

Inserting values of T in eqns A2.12, A2.14 and A2.16 yields values for the equilibrium constants, $K_{Al_2O_3}$, K_{MgO} and $K_{MgAl_2O_4}$. Inserting these values together with values for a_{Al} and a_{Mg} in eqns A2.13, A2.15 and A2.17 yields equilibrium oxygen activities.

A2.2.1.2 Example of Calculations

In this example, equilibrium oxygen activities at 500°C (773 K) are calculated for Al_2O_3, MgO and $MgAl_2O_4$ as functions of alloy composition.

For Al_2O_3:
From eqn A.2.12: $KAl_2O_3 = 2.8 \times 10^{65}$

Substituting for $K_{Al_2O_3}$ in eqn A2.13: $a_{o_2} = \dfrac{1}{2.8 \times 10^{65} \times (a_{Al})^{\frac{4}{3}}}$ $\qquad (A2.18)$

For MgO:
From eqn 2.14: $K_{MgO} = 3.1 \times 10^{70}$

Substituting for K_{MgO} in eqn A2.15: $a_{o_2} = \dfrac{1}{3.1 \times 10^{70} \times (a_{Mg})^{2}}$ $\qquad (A2.19)$

Table A2.1 Oxygen Activities in Equilibrium with Al_2O_3, MgO and $MgAl_2O_4$ on Aluminium - Magnesium Alloys at 500°C as Functions of Alloy Composition

mass% mg	X_{Mg}	a_{Mg}^*	a_{Al}^*	$a_{O_2} \times 10^{69}$		
				MgO	Al_2O_3	$MgAl_2O_4$
0.2	0.0025	0.0055	0.9975	1140	3531	112
0.5	0.005	0.011	0.9950	285	3543	79.1
0.9	0.01	0.022	0.990	71.3	3567	56.4
1.8	0.02	0.044	0.980	17.8	3616	40.2
2.7	0.03	0.066	0.970	7.9	3666	33.2
3.6	0.04	0.088	0.960	4.5	3717	30.0
4.5	0.05	0.110	0.950	2.9	3769	26.2
5.4	0.06	0.131	0.940	2.0	3822	24.3
6.3	0.07	0.152	0.931	1.5	3872	22.8
7.2	0.08	0.172	0.923	1.2	3916	21.6
8.2	0.09	0.190	0.916	0.9	3957	20.6
9.1	0.10	0.210	0.910	0.8	3992	19.8

*Brown and Platt[8]

For $MgAl_2O_4$:
From eqn A2.16: $K_{MgAl_2O_4} = 1.14 \times 10^{68}$

Substituting for $K_{MgAl_2O_4}$ in eqn A2.17: $a_{O_2} = \dfrac{1}{1.14 \times 10^{68} \times a_{Al} \times (a_{Mg})^{1/2}}$ (A2.20)

Brown and Pratt[8] give values for activities of aluminium, a_{Al}, and magnesium, a_{Mg}, for the corresponding mole fractions, X_{Al} and X_{Mg}. Inserting values for 500°C for compositions in the range of interest yields the equilibrium oxygen activities as functions of alloy composition given in Table A2.1 and plotted in Figure 8.11 in Chapter 8.

A2.3 FORMATION OF OXIDES ON ALUMINIUM-LITHIUM ALLOYS

Equilibrium oxygen activities for Al_2O_3, Li_2O and $Li_2Al_2O_4$ cannot be calculated as functions of alloy composition because there is no information on activities of lithium and of aluminium in the α phase of the binary system. However, to facilitate discussion in Chapter 8, Section 8.4.3 it is useful to compare values for the free energies of formation of the oxide Li_2O, the spinel, $Li_2Al_2O_4$ and the hydroxide, LiOH with their counterparts in the aluminium-magnesium system, MgO, $MgAl_2O_4$ and $Mg(OH)_2$. Eqns A2.2, A2.4, A2.6, A2.7, A2.9 and A2.11 yield the values given for 400, 500 and 600°C in Table 8.6.

A2.4 HYDRATION OF MAGNESIUM AND LITHIUM OXIDES

Magnesium Oxide

Assuming unit activities for MgO and $Mg(OH)_2$, application of the Van't Hoff isobar to the reaction: $MgO(s) + H_2O(g) = Mg(OH)_2(s)$

yields: $\Delta G_{(T,p)} = \Delta G_5^{\ominus} - RT \ln a_{H_2O}$

$$= -46024 + 100\, T - RT \ln a_{H_2O} \text{ Joules (from eqn A2.5)} \qquad (A2.21)$$

where: $\Delta G_{(T,p)}$ is the Gibbs free energy change for the reaction at an arbitrary activity of water vapour, a_{H_2O}.

Lithium Oxide

Assuming unit activities for Li_2O and LiOH, application of the Van't Hoff isobar to the reaction: $Li_2O(s) + H_2O(g) = 2LiOH(s)$

yields: $\Delta G_{(T,p)} = \Delta G_{10}^{\ominus} - RT \ln a_{H_2O}$

$$= -146189 + 108.7\, T - RT \ln a_{H_2O} \text{ Joules (from eqn A2.10)} \qquad (A2.22)$$

where: $\Delta G_{(T,p)}$ is the Gibbs free energy change for the reaction at an arbitrary activity of water vapour, a_{H_2O}.

Equations A2.21 and A2.22 are used to produce Table 8.4 in Chapter 8.

A2.5 REFERENCES

1. O. Kubaschewski and C.B. Alcock, *Metallurgical Thermochemistry*, Pergamon Press, New York, 1979.
2. R.L. Altman, *J. Phys. Chem.*, **67**, 1963, 366.
3. E.G. King, *J. Phys. Chem.*, **59**, 1955, 218.
4. K.K. Kelley, *Bull. U.S. Bur. Mines*, (447), 1950.
5. J.P. Coughlin, *J. Amer. Chem. Soc.*, **78**, 1956, 5168 and **79**, 1957, 2397.
6. E.G. King, *J. Amer. Chem. Soc.*, **76**, 1954, 5849, **77**, 1955, 3189 and **80**, 1965, 1799.
7. H.J. Johnston and T.W. Bauer, *J. Amer. Chem. Soc.*, **73**, 1951, 1119.
8. J. Brown and N. Platt, *Met. Trans. B*, **1**, 1970, 2743.

Subject Index